9/7 ELECTRONIC

ELECTRONIC ASSEMBLY

ELECTRONIC ASSEMBLY

Soft Soldering and Wire Wrapping

Gerry Herrick

Prentice Hall
Englewood Cliffs, New Jersey 07632

Library of Congress Cataloging-in-Publication Data

Herrick, Gerry.
Electronic assembly : soft soldering and wire wrapping / Gerry Herrick.
p. cm.
Includes bibliographical references and index.
ISBN 0-13-248766-7
1. Electronic apparatus and appliances—Design and construction.
2. Solder and soldering. I. Title.
TK7870.H445 1992
621.381—dc20 91-15683
CIP

Editorial/production supervision and interior design: *Harriet Tellem*
Cover design: *Wanda Lubelska Designs*
Prepress buyer: *Mary Elizabeth McCartney*
Manufacturing buyer: *Susan Brunke*
Acquisitions editor: *George Kuredjian*
Editorial Assistant: *Barbara Alfieri*

A Simon & Schuster Company
Englewood Cliffs, New Jersey 07632

The publisher offers discounts on this book when ordered in bulk quantities. For more information, write:

Special Sales/Professional Marketing
Prentice-Hall, Inc.
Professional & Technical Reference Division
Englewood Cliffs, New Jersey 07632

Printed in the United States of America

10 9 8 7 6 5 4 3 2 1

ISBN 0-13-248766-7

Prentice-Hall International (UK) Limited, *London*
Prentice-Hall of Australia Pty. Limited, *Sydney*
Prentice-Hall Canada, Inc., *Toronto*
Prentice-Hall Hispanoamericana, S.A., *Mexico*
Prentice-Hall of India Private Limited, *New Delhi*
Prentice-Hall of Japan, Inc., *Tokyo*
Simon & Schuster Asia Pte. Ltd., *Singapore*
Editora Prentice-Hall do Brasil, Ltda., *Rio de Janeiro*

Dedication

To my husband, Clyde, who encouraged and inspired me to write this book.

To my children, Lara and Johnny, for their patience and support during the many hours of research and writing.

To my students, whose quest for learning is an inspiration to me and the reason for remaining in the field of education.

Contents

Preface

Soldering, the joining of two metals, was introduced in ancient Egypt around 35 B.C. The basic principles of this ancient craft are still used in industrial processes.

Soft soldering has been used in the electronic industry for well over 60 years. It comprises approximately 50% of all soldering performed today. Without the use of solder, a strong, permanent, metallic and electrical bond would not be possible. A good electrical and mechanical bond is necessary to maintain reliable equipment operation. Faulty solder joints are one of the major causes of equipment failure, a problem that cannot be tolerated in today's highly technical and electronic-dependent society.

Although the same basic principles of soldering still apply today, there has been great improvement in the process and materials used. Today's soldering is far from simple. It requires care, experience, and knowledge of fundamentals. Prior to the printed circuit board (PCB), all soldering was performed by hand. After the development of the PCB, soldering was either performed by hand or the boards were hand dipped in a solder bath, called a solder pot. This process proved to be slow and tedious. Since the components on a PCB are mounted on a flat plane, automated or machine soldering became possible. Machine soldering, also known as wave soldering, makes over 1000 connections in a few seconds. If properly controlled, wave soldering is more economical and produces boards with fewer rejections than the hand method.

In some situations, such as in connecting electrical wires, cables, and harnesses, automated soldering cannot be used. In these situations hand soldering or mechanical connections are made. Hand soldering is also used for making repairs of faulty joints, shorts, bridges, and other defects found during inspection. It is also used for rework, that is, making electrical changes on the PCB after components have been mounted and soldered.

Whatever method is used, the importance of high-standard workmanship cannot be overemphasized. Today, in the electronic industry, soldering is a critical skill. Engineers, electronic technicians, and electronic assemblers will need to know how to solder and how to do it well. This skill is necessary at the production stage, for building prototype projects, and for rework and repair.

The emphasis of this book is placed on showing how to make reliable, high-quality hand-solder and wire-wrap connections. Other methods of soldering and electrical assembly are also briefly discussed.

ACKNOWLEDGMENTS

Thank you to Roger Trovato of Dover-Soltec, and to Jayne Stafstrom of dataCon, Inc., for all of their technical assistance with this book. Also, thank you to Jiann Wany and Alwin Narayan for all of their help in processing the photographs.

My sincere appreciation to the artist for this book, Robert Mosher, for his technical assistance and his graphics expertise, which were essential in the development of the book.

Chapter 1

Electronic Assembly

1.1 INTRODUCTION

The electronic assembler performs the task of properly installing parts and components onto printed-circuit (PC) boards and into electrical equipment. There are several categories of assemblers:

1. Assemblers that are responsible for installing only the mechanical hardware, such as nuts, bolts, and handles.
2. Assemblers that install the electrical components onto printed-circuit boards and perform the wire wrapping or soldering. They are usually utilized where single models (prototypes) of a product are produced.
3. Assemblers that solder the components or wire wrap the sockets after they are inserted into the printed-circuit board. They are usually involved in medium-sized production runs, where it is not economical to utilize automated part-insertion equipment.
4. Assemblers that make corrections to soldering connections missed in automatic soldering are usually used in large production runs. In such runs the component insertion and soldering process is usually automated. The parts are inserted into the PC boards by a robot and then fed through an automatic soldering process.

5. Assemblers that make corrections on finished boards and finished products. Regardless of the assembly, soldering, and/or wirewrap process, there is usually a requirement for an assembler to correct soldering errors and replace faulty components that are found during testing. These workers are usually the most skilled assembly personnel. This type of assembler worker is usually utilized in the final product test or in large production runs.

Obviously, each assembler performs an important and necessary function in the production of electronic equipment.

The purpose of this text is to act as a guide for the professional who is responsible for assembly and rework and the instructor who teaches assembly processes, and as a training aid for the assembler who must perform the soldering and wirewrap functions in either first assembly, correction of automatic assembly errors, or board rework. This chapter introduces the assembler to the terminology and safety precautions that are necessary to correctly and safely perform the assembly job.

1.2 TERMINOLOGY OF THE TRADE

An **electric circuit** is defined as the interconnection of a number of devices in one or more closed paths to perform a desired electrical function. It provides a path for electrons to travel to the load and return to their starting point. If the circuit is **open** (not complete), electrons will not flow. A poor soldering connection, a broken or loose wire, a damaged device, or a switch in the open position stops electron flow. Electron flow is usually called **current flow.**

A **complete electrical circuit** consists of four basic units:

1. Conductor or current path
2. Load
3. Energy source (voltage)
4. Control device

A **conductor** is any of the materials used to transport energy from one place to another. A conductor can be copper wires, fiber-optic cables, or air in the case of radiated energy. We are concerned only with copper conductors in this text.

A **load** in a circuit can be any of many components or devices, such as resistors, light bulbs, speakers, motors, computer terminals, printers, or a personal computer.

An **energy source** can be a battery, a generator, an amplifier, a digital computer or any other device that has an output of information in the form of a voltage.

Control devices are circuit components such as knife switches, circuit breakers, or a pullout fuse block used for connecting and disconnecting the power source to a circuit or system.

Voltage in a circuit acts as a pressure and causes the current to flow through the various components of the circuit. Voltage can be direct current (dc), like that of a battery, or alternating current (ac), like that used in homes and factories. The name for voltage is volts and the symbol is E or V.

Current flow is the movement or flow of electrons through a circuit that is caused by the voltage applied to the circuit. When a voltage is applied to a complete circuit from a source (such as a battery), electrons move through the conductors of the circuit (energy transfer) to the load at the receiving end of the conductors. The desire is to transport the electrical signals along the conductors (wires) and have them arrive at their destination to do some work; for example, to drive a speaker in a radio. The unit of current is the **ampere** and the symbol is A.

Resistance is the property of an electrical circuit that limits the current. The resistance of an electrical circuit can be compared to friction in a mechanical device, where the frictional drag on a body, limits the speed of the body and produces heat. The unit of resistance is the ***ohm,*** which is symbolized by the Greek letter omega (Ω). The ohms value of a resistor is usually specified by a color code on the resistor. The color code uses ten colors to represent numbers. Table 1–1 gives the colors and the numbers that they represent.

The color code is used to specify the ohms value of the resistor as follows: the first band on the resistor is the first number, the second band is the second number, and the third band is the number of zeros to be added. For example, if a resistor has color bands of red, violet, and brown, then the value is 2, 7, and add 1 zero, or 270 ohms.

Most resistors have a fourth band that is silver or gold (Fig. 1–1a). This band gives the tolerance of the resistor as 5% for gold and 10% for silver. This is because resistors, like most other things, seldom need to be made exact. For example, a brown-, black-, brown-, and gold-coded resistor would have a value of 100 ohms ± 5%. In other words, the resistor could have a value between 95 and 105 ohms and still

TABLE 1–1 COLOR CODE USED TO IDENTIFY THE VALUE OF RESISTORS AND TOLERANCES.

Color	Significant figure	Decimal multiplier	Resistance tolerance in percent
Black	0	1	±0
Brown	1	10	±1
Red	2	100	±2
Orange	3	1,000	±3
Yellow	4	10,000	±4
Green	5	100,000	±5
Blue	6	1,000,000	±6
Violet	7	10,000,000	±7
Gray	8	100,000,000	±8
White	9	1,000,000,000	±9
Gold	—	0.1	±5
Silver	—	0.01	±10
No color	—		±20

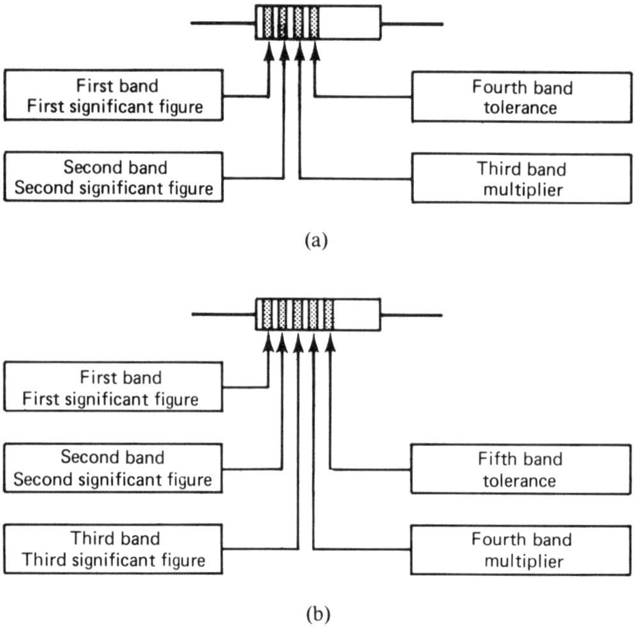

Figure 1–1 Color code for resistors. (a) Four-band code for resistors with a ±1% to ±10% tolerance. (b) Five-band code for resistors with a ±1% to ±10% tolerance.

be considered good. If the code band had been silver, the resistor's value might have been between 90 ohms and 110 ohms.

The manufacturing of resistors is becoming so competitive that you will seldom see 10% resistors. In fact, many manufacturers are producing five-band resistors for greater accuracy. These are coded where first band = first number, second band = second number, third band = third number, fourth band = number of zeros, and fifth band = tolerance. The tolerance might be any color. For example, brown would be 1% (Fig. 1–1b). A five-band resistor with the colors of orange, blue, green, red, and brown would have a value of 3650 Ω ± 1%.

The senior assembler must know the color code to resistors and how to recognize resistors of different wattage ratings, as well as many other electronic parts.

Power in a circuit is the rate of doing work. In electrical circuits, power is measured by units called **watts.** Electrical components and devices usually have a power or wattage rating. For example, a resistor may have a $\frac{1}{8}$-, $\frac{1}{4}$-, $\frac{1}{2}$-, or 1-watt or even higher rating. The size of the resistor determines its wattage rating. An assembler must make sure that the resistor has both the specified ohms value and correct wattage rating (Fig. 1–2).

1.3 GROUND

Ground, or earth as the British say, is thought of as 0 volts. It is used as a reference point for measuring voltage. The term ground is also used to mean the metal case or

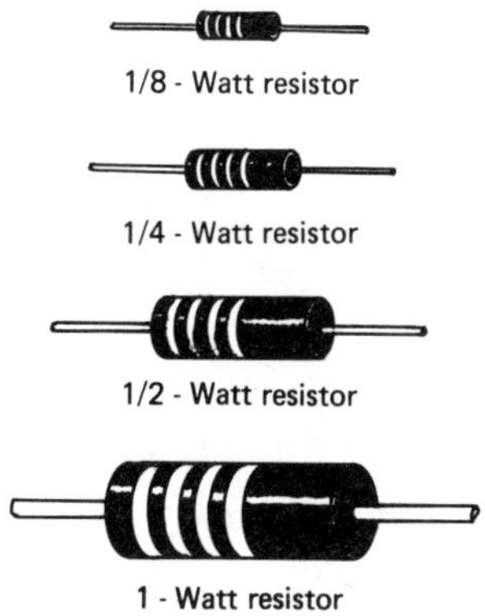

Figure 1–2 Example of resistors of different wattage ratings: $\frac{1}{8}$, $\frac{1}{4}$, $\frac{1}{2}$, and 1 watt.

chassis of a piece of electronic equipment. U.S. government safety codes specify that all electrical equipment must be electrically connected to ground (grounded) to prevent an electrical potential (voltage) from occurring between the equipment and ground and between equipment. This is a safety precaution designed to protect both people and equipment. The symbol for ground is shown in Fig. 1–3.

The grounding of electrical equipment is usually accomplished through the power plug. For 120-volt connections the center prong of the three-prong plug is ground (Fig. 1–4). The insulation color of the ground wire within a power cord to equipment is usually green.

Semiconductor devices of the MOS type can be destroyed by static voltage during handling and installing. Therefore, every person from the shipping clerk to the assembler must know the proper safe handling techniques for these devices. (Section 3.3 gives a detailed summary of these handling techniques.)

1.4 INTRODUCTION TO SAFETY

The electronic devices, equipment, and tools that are used by workers in the electronics industry can be hazardous if not properly used. Good safety habits must be followed at all times for the protection of workers and others. Many companies have safety training programs for their employees in an attempt to minimize accidents and injuries. However, in spite of the best efforts of management, many accidents occur each year. These accidents cause serious injuries, resulting in lost work time and even deaths. **Safety is the responsibility of everyone in the workplace.**

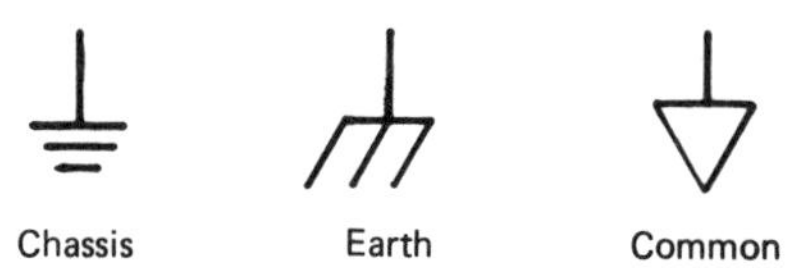

Figure 1–3 Symbols for ground.

Figure 1-4 Example of a grounded ac power plug.

1.5 ELECTRICAL SHOCK

A major hazard and one of the most common for people who work with and on electrical equipment is **electrical shock.** Electrical shock is caused by the passing of electrical current through the body. The amount of current flow is directly related to the applied voltage. The higher the voltage is the greater the current flow and the more serious the shock.

However, you should not get the idea that low voltages will not cause shock; they can under the right conditions. Resistance limits current. Therefore, when resistance decreases, current will increase, and vice versa. The overall resistance of the human body can vary from a low value to a very high value depending on physical conditions. Dry skin presents a high body resistance of several hundred thousand ohms to a voltage, so only a small current will flow through the body with, say, a voltage of 120 volts. However, wet or damp skin results in a body resistance of less than a 1000 ohms. In this case, even a low voltage could be fatal since a large current could pass through the body. The very low voltage on which a telephone operates could be fatal if the person were in a bathtub or swimming pool. A big factor in the severity of the shock from a voltage could be the type of shoes that a person is wearing

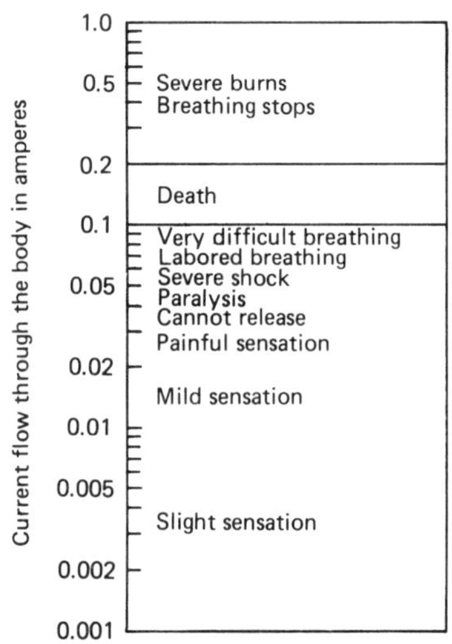

Figure 1-5 Degrees of electrical shock and accompanying body sensations.

and the type of floor on which he or she is standing when the connection to the voltage is made.

The chart in Fig. 1–5 shows the effects on the human body of different levels of current.

For general shop safety rules (and for tests), refer to Appendix J.

It is recommended that workers review these rules for their own safety. Instructors should have their students review these rules and take the safety test before they undertake any activity in the shop or laboratory.

TEST

1. What is the job of the electronic assembler?
2. Who is responsible for safety in the shop or laboratory?
3. Why is it important that every soldering connection and every wire-wrap connection be perfect?
4. What is the purpose of the ground on electrical equipment?
5. Define current, voltage, and resistance as they apply to an electrical circuit.

Chapter 2

Tools of the Trade

2.1 INTRODUCTION

Electronic assembly is the process of building functional electrical units by properly mounting and soldering and/or wire wrapping connectors, components, and wires into the units. This requires the ability to recognize the different components, materials, and equipment used in the assembly. An understanding of the operation of these components and materials is desirable, but not necessary. In addition to these skills, the assembler must be familiar with the hand tools used to complete the task.

Hand tools are used often in all aspects of electronic fabrication. A knowledge of their proper use simplifies work by enabling the user to select the right tool for a specific application and to use the tool safely. The most commonly used tools are identified and their applications are discussed in this chapter.

2.2 PLIERS

There are many types and styles of pliers for general-purpose and special applications. The following are those most often used in electronic assembly.

Long needle nose pliers, also known as long nose or needle nose, come in various sizes for different applications (Fig. 2–1). They have long slender jaws with a fine needle point. Long needle nose pliers are delicate and should not have much force applied to them. They should never be used on heavy wire or for any heavy-duty work. They are used for precision bending of fine wire, for reaching into confined areas, and for guiding leads through high-density wire areas.

Slim nose pliers have very fine tips, but the base of the slim jaws is heavier than long needle nose for added strength (Fig. 2–2). They are used for deeper reach into hard to get to areas, holding and handling small wire and parts, and crimping wires on terminal posts. These pliers are ideal for production line work. Slim nose pliers, like long needle nose pliers, are not meant for heavy-duty work. Their fine, delicate jaws can be easily damaged if improperly used.

Curved nose pliers are similar to long nose pliers, except they are bent to approximately a 45° to 70° angle, depending on the application (Fig. 2–3). They are designed for forming fine wire, working around obstructions, and getting into hard to reach spaces. These pliers, when used, provide full view of the work area and are particularly suited for PC board rework. Curved nose pliers are also delicate and should not have excessive force applied to them.

Transverse-end cutting pliers have a firm jaw much like slim-nose pliers, but they also have cutting knives at the tip (Fig. 2–4). These pliers can be utilized for cutting wires in hard to get at places. The end cutters are not designed for heavy-duty work.

Diagonal cutting pliers, also known as *dikes,* have a short nose and come in various sizes for different applications (Fig. 2–5). They are made in two basic styles: flush (or full flush) and semiflush. The full-flush pliers are designed to reduce the amount of pinch left on the cut wire. Full-flush cutters are recommended for cutting no. 22 gauge wire or smaller, for trimming small leads, and for other delicate jobs. The semiflush cutters are used for cutting nos. 18 to 20 gauge wire, trimming plastic, and terminal leads and for general-purpose assembly line production work. They are not designed for stripping insulation from wires. They can nick the wire and cause the wire to weaken or break.

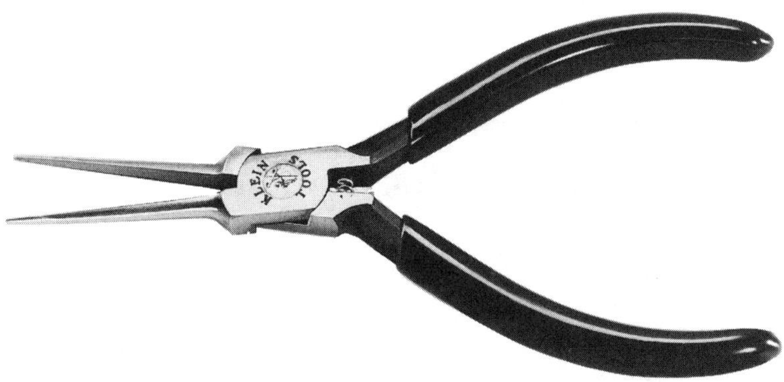

Figure 2–1 Long needle nose pliers. (Courtesy of Klein Tools, Inc.)

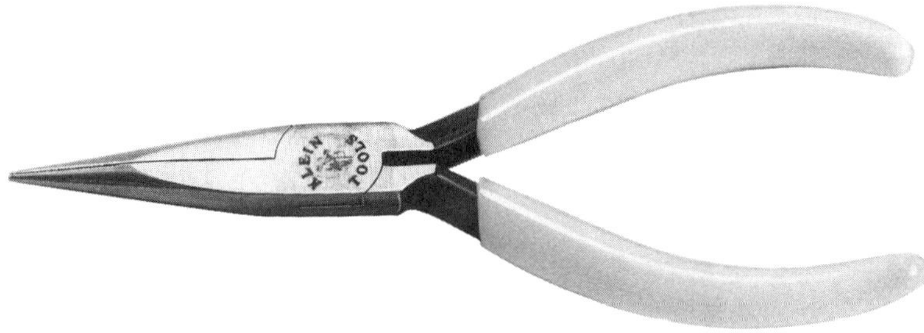

Figure 2-2 Slim long-nose pliers. (Courtesy of Klein Tools, Inc.)

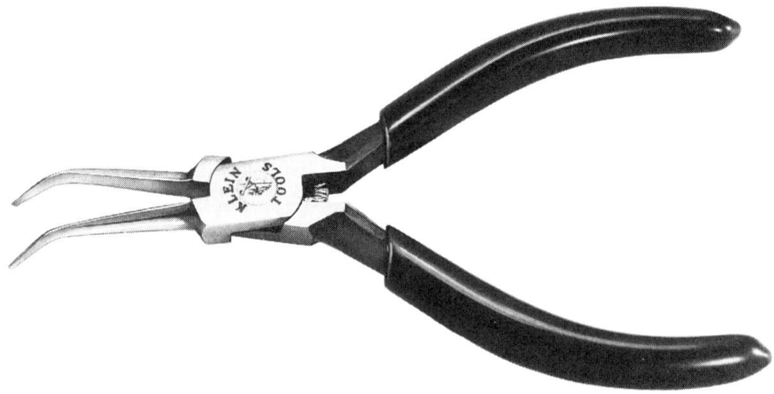

Figure 2-3 Curved nose pliers. (Courtesy of Klein Tools, Inc.)

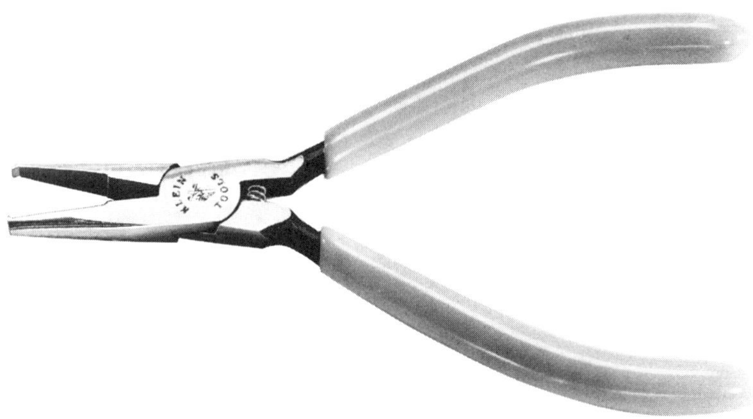

Figure 2-4 Transverse-end cutting pliers. (Courtesy of Klein Tools, Inc.)

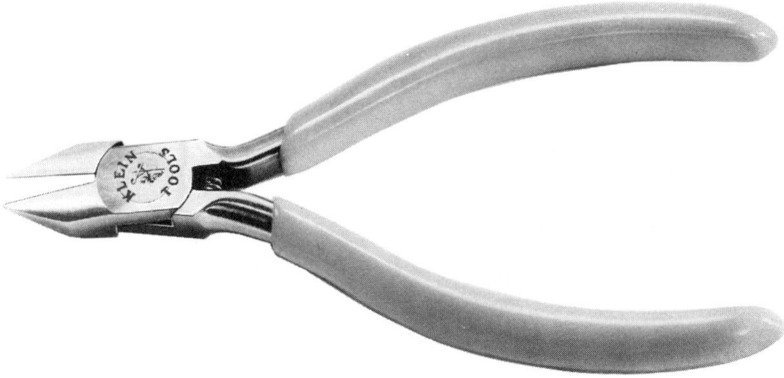

Figure 2–5 Diagonal cutting pliers. (Courtesy of Klein Tools, Inc.)

Short nose, heavy-duty diagonal pliers have a short nose and a heavy jaw with side cutting edges (Fig. 2–6). The large handles and short cutting edges make this a powerful cutting tool ideal for cutting heavy wire, PC boards, and plastics and for other cutting in general assembly use.

The aforementioned pliers are often available with spring-loaded handles for ease of use. Figure 2–7 illustrates two of these types of pliers.

2.3 WIRE STRIPPERS

Wire strippers are used to remove insulation from wires and come in many styles. Most strippers resemble pliers, but they differ by having cutting edges notched to different wire sizes (Fig. 2–8a). This style of stripper is not recommended for electronic assembly, because it tends to nick the wire.

Dual-blade wire strippers, however, are more reliable (Fig. 2–8b). They allow the insulation to be stripped off without cutting or fraying the wire. Their automatic operation holds the jaws open after stripping to permit removal of wire and prevent crushing. A single squeeze of the handle strips the wire clean. Dual-blade wire strip-

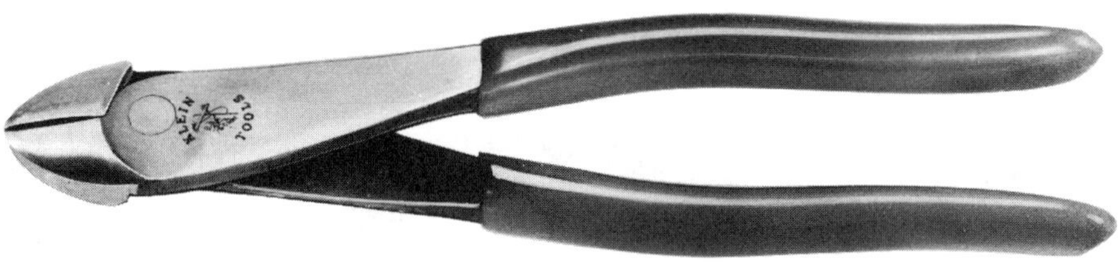

Figure 2–6 Short-nose, heavy-duty pliers. (Courtesy of Klein Tools, Inc.)

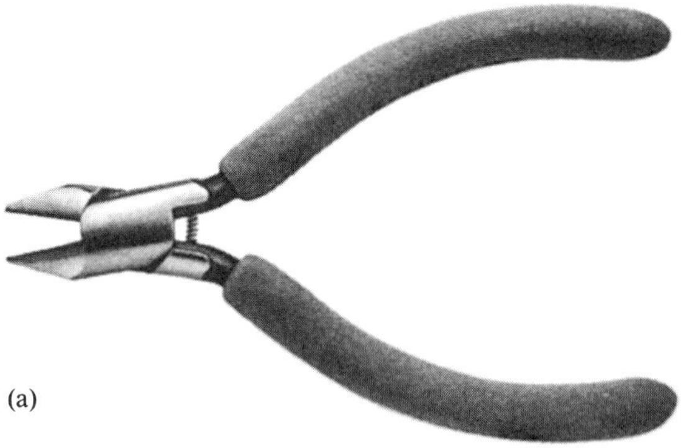

(a)

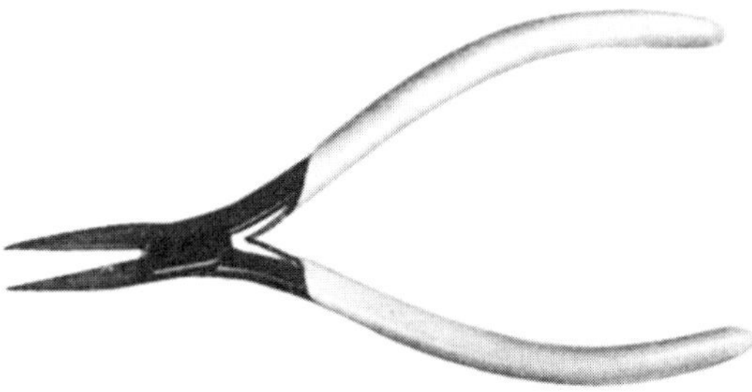

(b)

Figure 2–7 Spring-loaded diagonal and needle nose pliers. (Courtesy of Klein Tools Inc.)

pers have interchangeable blades that can strip wires from sizes 10 to 30 gauge. Care must be taken to insert the wire into the notch corresponding to the wire size.

Thermal wire strippers (Fig. 2–8c) remove the insulation by heat. Although this type of stripper will not nick the wire, it cannot be used with all types of insulation.

2.4 SCREWDRIVERS

Screwdrivers are one of the most commonly used tools. They consist of two parts: (1) blade and shaft and (2) the handle. The most common designs are the flat blade and the Phillips. They come in many sizes because different-sized screws require different-sized screwdrivers. To complete a task properly requires selecting the appropriate size and design type of screwdriver.

Flat-blade screwdrivers are used to tighten or loosen a screw with a slotted head (Fig. 2–9). The blade of a flat-blade screwdriver has a chisel-shaped head that mates with straight slotted screws. To avoid any play, which could damage the screw head, use a properly fitting screwdriver at all times.

The *Phillips screwdriver* has a notched tip (Fig. 2–10). It is designed to tighten

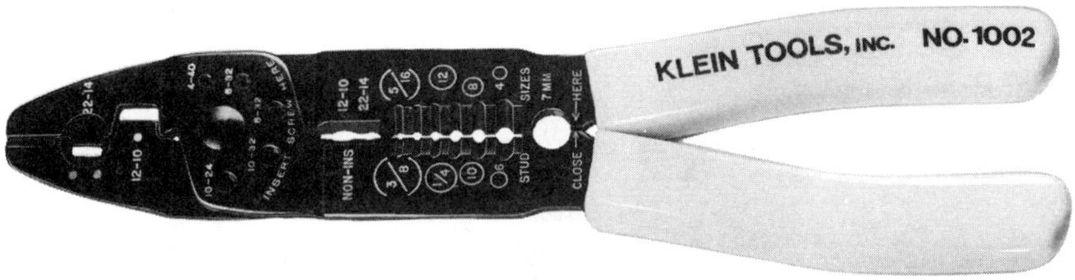

(a)

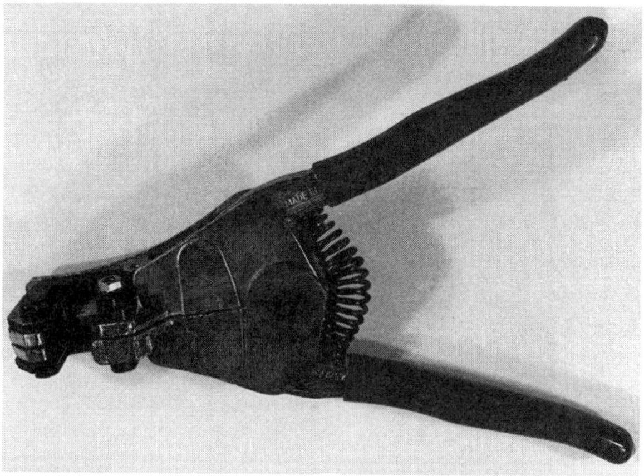

(b)

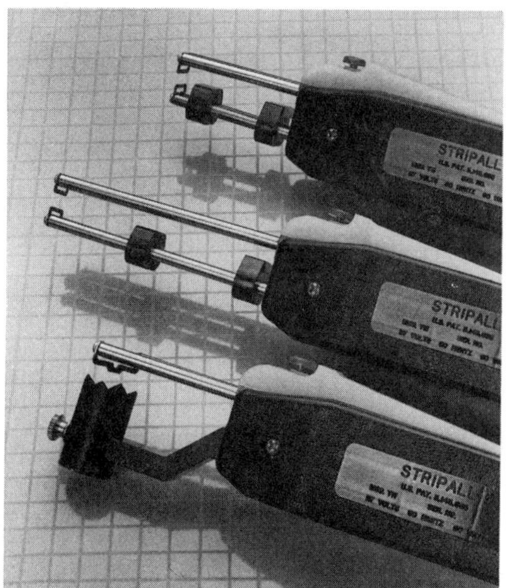

(c)

Figure 2–8 Types of wire strippers. (a) Combination-style wire strippers. (Courtesy of Klein Tools, Inc.) (b) Dual-blade wire strippers. (c) Thermal wire strippers. (Courtesy of Teledyne-Kinetics)

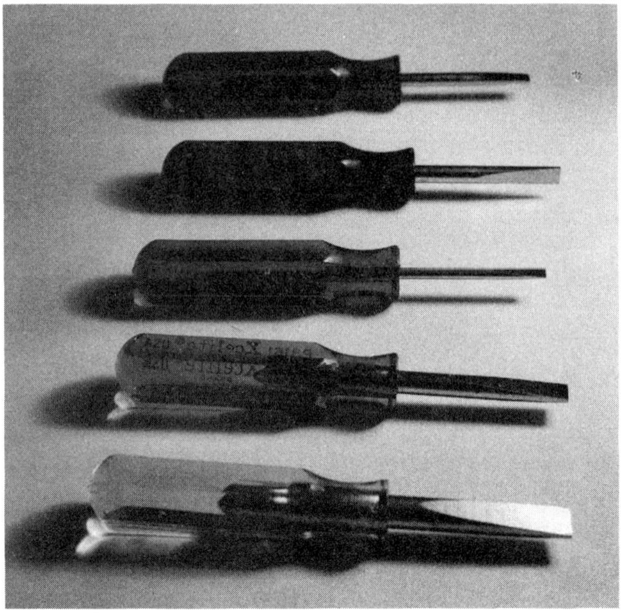

Figure 2–9 Flat-blade screwdrivers. (Courtesy of Cooper Tools, Inc.)

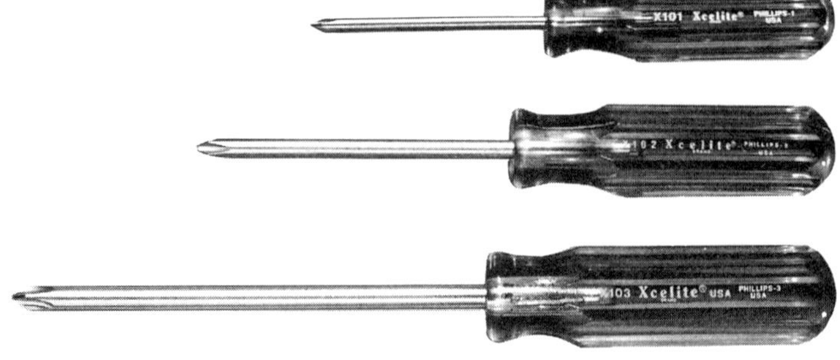

Figure 2–10 Phillips screwdrivers. (Courtesy of Cooper Tools, Inc.)

or loosen screws with star-shaped (cross-head) openings. Like the flat-blade screwdriver, it comes in varying sizes. Care should be used in selecting the appropriate-sized screwdriver to fit the notched hole. The most common error in using the Phillips screwdriver is to select one that is too small and thereby ruin the screw head. There are other special types of screwdrivers. Some equipment manufacturers are "making up" screw types that require the purchase of their screwdrivers.

2.5 NUT DRIVERS

A nut driver is a fixed socket wrench with a screwdriver-type handle. It is used to tighten or remove nuts (Fig. 2–11) and is particularly useful to remove nuts located in hard to reach places. Nut drivers usually come in sets color coded for size. The most often used sizes range from $\frac{3}{16}$ to $\frac{3}{8}$ inch. The thickness of the socket walls varies with the size of the socket. Nut drivers in metric sizes are available. Although the shaft varies in length, the most often used length is approximately 4 inches. Nut drivers can be purchased individually to replace those that are misplaced.

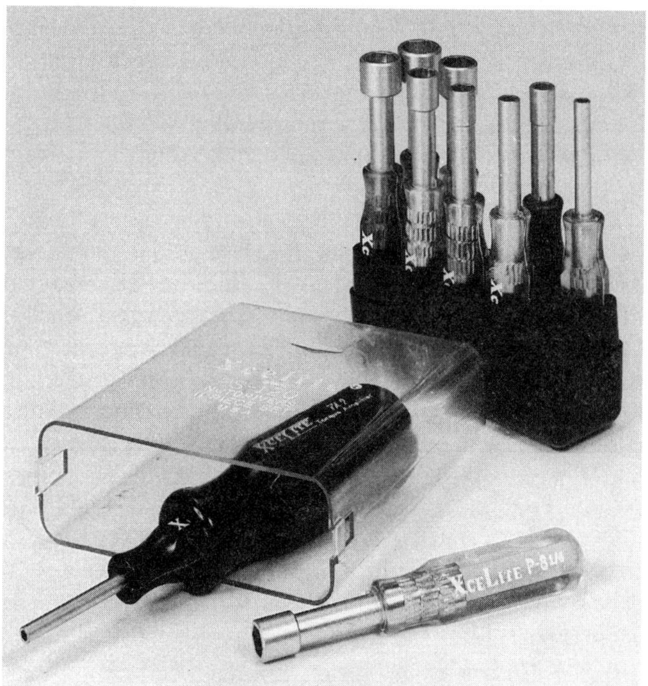

Figure 2–11 Nut drivers. (Courtesy of Cooper Tools, Inc.)

2.6 SOLDERING TOOLS

A *soldering iron* is the most important tool of the assembly process, and using the soldering iron properly is the most demanding skill of an electronics assembler. To solder correctly, the proper soldering iron must be selected to fit the job. Incorrect use or the wrong style of iron can cause faulty soldering connections, resulting in unit failure. The purpose of the soldering iron is to transfer enough heat to the parts you want to join to melt the solder without damage to wires, terminals, and components.

There are two basic type of soldering irons: guns and pencils. **Soldering guns should never be used for electronic assembly.** Pencil-style soldering irons may be individual (without a soldering station) and operate directly on 120 V ac. The hand-held pencil with a soldering station operates on a low voltage that is transformed from the 120-V ac line by the soldering station (Fig. 2–12). All soldering irons must have a connection to ground. For an explanation of grounding, see Section 1.3.

Pencil soldering irons have removable tips for replacement to allow the use of different-shaped tips for a variety of soldering applications (Fig. 2–13). The tips are available with different temperature ratings. For general soldering, a 600°F to 800°F tip is recommended.

Soldering irons are rated in watts, which is an indication of how quickly they will transfer heat. For soldering most wires, terminals, or PC boards, a 40- to 50-watt iron or less will transfer enough heat in a short period of time to obtain a good solder joint without damage to heat-sensitive devices. For soldering of integrated circuit (IC) terminals or very small printed-circuit pads, a $22\frac{1}{2}$-watt iron is usually sufficient and will limit the possibility of overheating the IC or the PC pad.

Soldering stations come with grounded, line-isolating transformers to prevent line-voltage leaks and spikes that can damage semiconductor devices. They have low-voltage heating elements and temperature control capabilities.

Flame soldering with a *soldering torch* is used to heat workpieces that are too large to heat with a soldering iron. It can be used in plumbing, jewelry work, and the joining of large conductors. The work can be heated directly by the flame or with the use of a large tip. Care of the tip is the same for the torch as for the iron. This heating method is very rapid and should not be used on electronic equipment and PC boards. The intense heat of the flame will damage the board and equipment.

Some torch temperatures can be adjusted by controlling the gas to air flow or by using different parts of the flame. The tip or blue part of the flame is the hottest. Torches can be hooked directly to small gas cylinders or connected by a hose to larger gas cylinders or a natural gas line.

Some small torches are designed specifically for electronic soldering, but these torches are about the size of a marking pen and have a small tip. These miniature propane soldering torches are convenient for fieldwork where electric outlets are not available (Fig. 2–12c).

A *soldering aid* is a versatile tool that has applications in most soldering jobs (Fig. 2–14). One end is pointed and is used to remove excess rosin from soldered joints, to unwind wires, and to twist or hold wires.

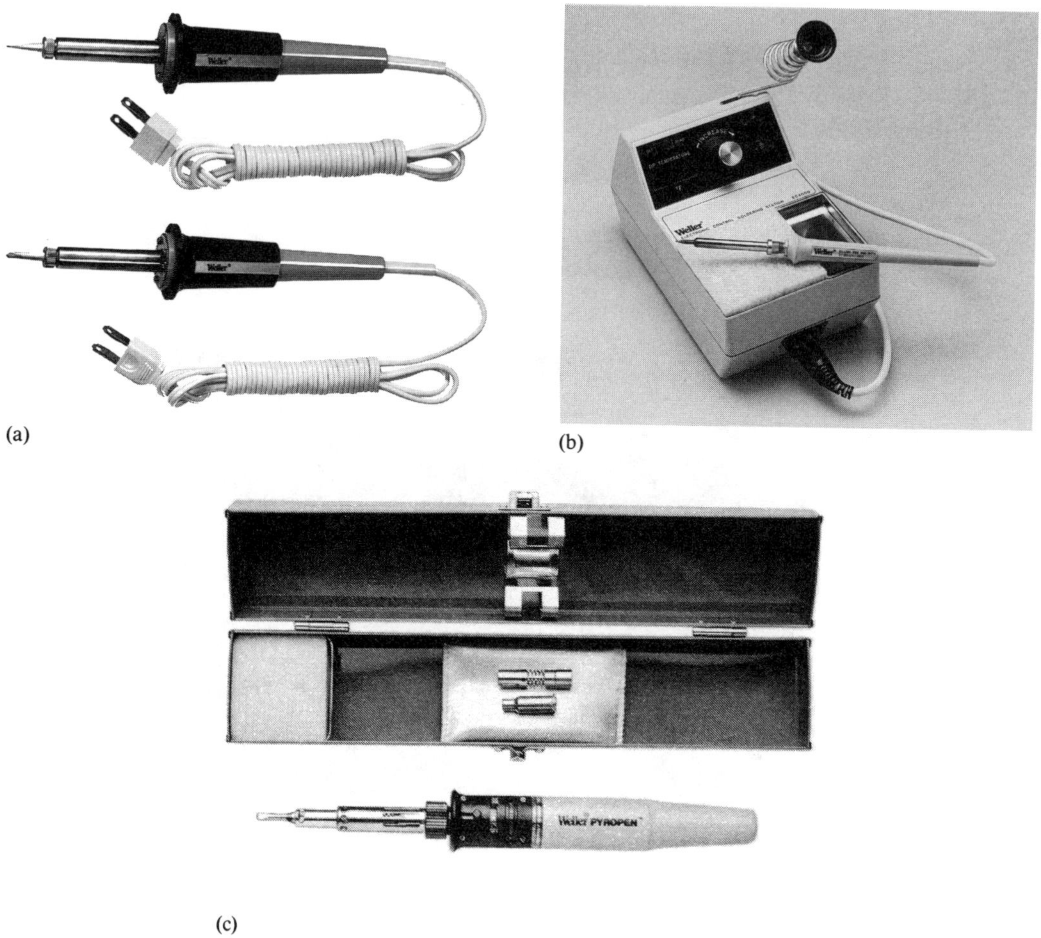

Figure 2–12 Pencil-style soldering irons. (Courtesy of Cooper Tools, Inc.) (a) Hand-held soldering iron. (b) Soldering station. (c) Weller Pyropen.

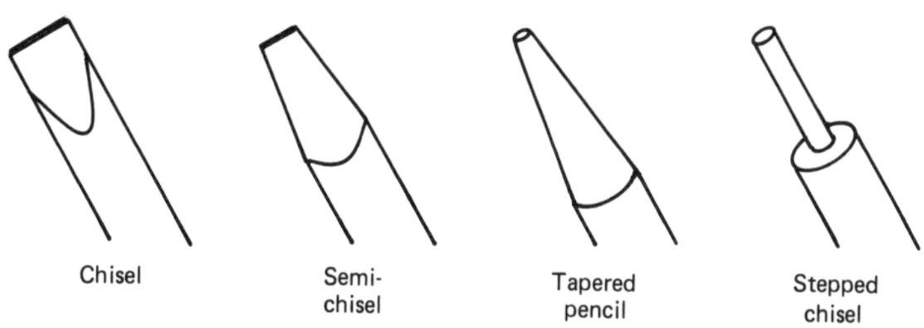

Figure 2–13 Samples of soldering iron tips.

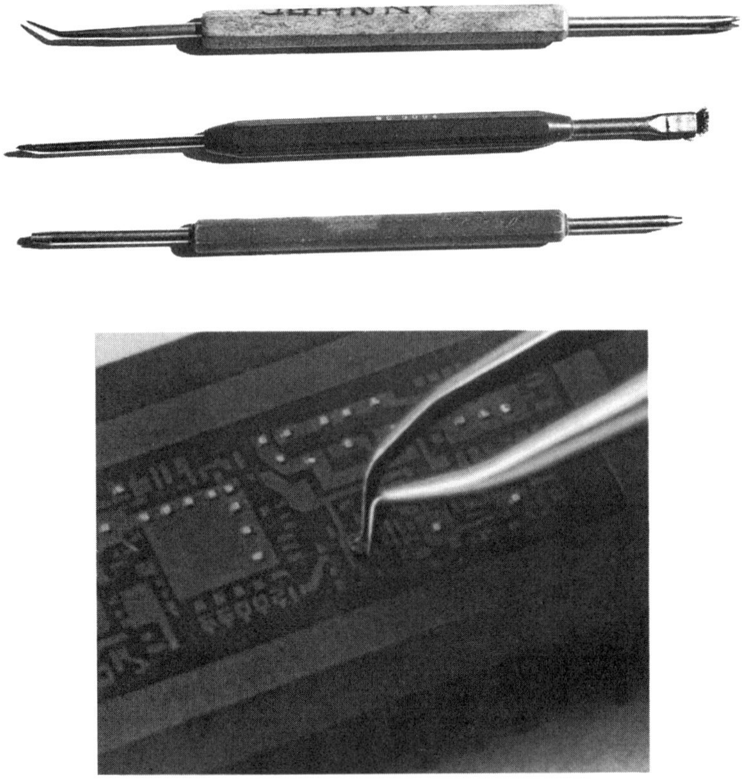

(b)

Figure 2-14 Soldering aids. (a) For standard leaded devices. (b) For surface-mounted devices. (Courtesy of Excelta Corp.)

The *antiwicking tool* works like self-closing pliers. When the handle is released, the jaws tighten, and the spring tension holds wires and components in place (Fig. 2–15). Antiwicking tools prevent solder from running under the insulation when tinning wire. Antiwicking tools can also be used as heat sinks, which prevent damage to heat-sensitive components by transferring heat to the metal clip.

A *desoldering device* is used to remove molten solder from soldered terminal connectors and mounted components (Fig. 2–16). Although this process should be avoided, sometimes it is necessary. The most popular desoldering device (solder sucker) uses a vacuum action to remove molten solder. The solder is heated as the Teflon tip of the solder sucker is applied to the molten solder. The solder sucker (Fig. 2–16a) uses a spring-loaded plunger to suck up the molten solder. Other methods of removing solder are rubber suction bulbs with a Teflon tip (Fig. 2–16b) and a desoldering wick, a braided copper material that uses capillary action to pull molten solder away from the heated terminal or electrical connection.

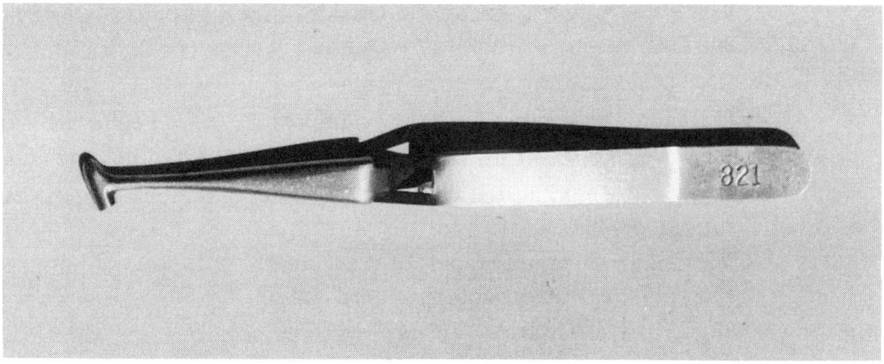

Figure 2–15 Antiwicking tool.

2.7 TOOL AND EQUIPMENT SAFETY

The following rules should always be followed when working with electrical or electronic equipment or performing work with electrical or mechanical parts.

1. Before working on electrical equipment, turn off all power and ground all high-voltage points. Make sure that the power cannot be restored accidentally.

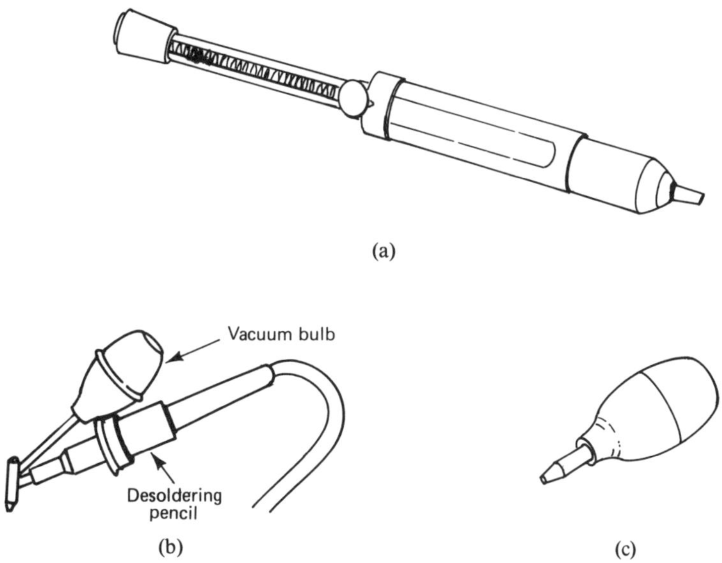

Figure 2–16 Desoldering tools. (a) Solder sucker. (b) Soldering iron with bulb. (c) Bulb type.

2. When you cut wires, keep the open side of the wire cutter away from your body and keep cut-off wire ends in your area.
3. Keep protective cover over knife blades.
4. Place tools and materials so they cannot slide, roll, or fall.
5. Never work on rotating machinery with someone standing too close.
6. Never use a box, barrel, chair, or any other makeshift support as a substitute for a ladder.
7. Replace or repair splintered, broken, rough, or loose tool handles before use. Use only files that are equipped with handles.
8. Shield dangerous parts of sharp-edged or pointed tools when you carry them from place to place.
9. Keep screwdriver blades sharp and even. (Do not use screwdrivers with broken or rounded points or broken handles).

HAND TOOL TEST

Match the name to the description by placing the letter of the description in the blank opposite the name of the appropriate tool.

Name of Tool

_____ 1. Soldering iron
_____ 2. Slim nose pliers
_____ 3. Solder sucker
_____ 4. Nut driver
_____ 5. Long needle nose pliers
_____ 6. Dual-blade wire strippers
_____ 7. Antiwicking tool
_____ 8. Diagonal cutting pliers
_____ 9. Phillips screwdriver
_____ 10. Soldering aid
_____ 11. Flat-blade screwdriver
_____ 12. Transverse-end cutting pliers
_____ 13. Curved nose pliers
_____ 14. Heavy-duty short nose plier

Hand Tool Description

A. A device used to prevent solder from running up under the insulation.
B. A device used to remove molten solder.

C. The cutter near the tip of these pliers helps cut wire in hard to get at places.

D. Delicate pliers used for bending fine wire.

E. Tool used for removing wire insulation.

F. A tool that is pointed at one end and used for twisting, holding, and unwinding wires.

G. A tool for cutting wires, also known as dikes.

H. Pliers that are bent to approximately a 45° to 70° angle.

I. Powerful cutting tool for cutting heavy wire.

J. Pliers designed to reach into deep, hard to get to areas.

K. Used to tighten or loosen screws with a slotted head.

L. A tool used to heat a terminal or electrical connection.

M. A fixed socket wrench used to quickly tighten or remove nuts or lugs.

N. A tool used to tighten or remove screws with star-shaped holes or cross-head openings.

Chapter 3

Parts and Materials

3.1 INTRODUCTION

Electronic components are functional parts connected together to form an assembled product. This chapter will present illustrations, symbols, and a brief description of the most common components and their function. For quick reference, refer to Appendix I for a listing of the most common components, their standard reference letter, and the unit of measure.

3.2 COMPONENTS, SYMBOLS, AND REFERENCE LETTERS

An **antenna** is an exposed conductor or the rods of a radio transmitter or receiver station used for the reception or transmission of radio waves from space (Fig. 3–1).

Batteries are a source of electrical energy. Batteries are constructed in many shapes and sizes. A battery is a dc voltage source, consisting of two or more cells (Fig. 3–2). It converts chemical energy into electrical energy. The reference letter for the battery is **B.**

Bulbs are light sources, which are made in various shapes and designs, depending on their function. Figure 3–3a is a neon lamp and Fig. 3–3b is an illumination

(a)

(b)

Figure 3–1 Antenna. (a) Television. (b) Symbol.

bulb. A neon lamp is a glow lamp that contains two elements and is filled with a neon gas. When a voltage of approximately 55 volts is applied to the lamp, the neon gas ionizes and glows orange. The illumination or incandescent bulb is the type of bulb most often used in homes for lighting and in flashlights and cars. The bulb uses a filament made of resistive material, usually a thin piece of tungsten, enclosed in an evacuated glass bulb (*vacuum*). Current flowing through the filament causes it to glow white hot, producing a brilliant light. The reference letters are **LP.**

A **capacitor** is a device consisting of two conducting surfaces separated by an insulating material called a dielectric: air, paper, mica, glass, plastic, film, or oil. Capacitors have the ability or *capacitance* to store an electrical charge. Some of its functions are to block current and permit the flow of current in amplifiers and to filter ripple voltagers in power supplies and store electrical charges, for example, in the high-voltage section of a TV. The amount of capacitance is related to the size of the plates, the distance between the plates, and the dielectric material being used. Capacitors are measured in farads (F). Typical values of capacitance are measured in microfarads (μF, 1×10^{-6}) and picofarads (pF, 1×10^{-12}).

Capacitors are generally grouped into three categories: paper foil or dry, electrolytic or wet, and air or variable (Fig. 3–4). The reference letter for a capacitor is **C.**

The electrolytic type of capacitor has the largest capacitance and must be placed in the circuit according to its polarity. Voltage polarity is indicated on one end of the capacitor. A band indicates the negative end. The positive end is sometimes indicated by a + symbol. Close inspection will show that the electrolytic capacitor is stamped out of a piece of aluminum, like a soft drink can. The metal end is **always** the negative end.

Coils or **inductors** are formed by winding a number of turns of wire around a form made of insulating material or an iron core. By means of a magnetic field, a coil

(a)

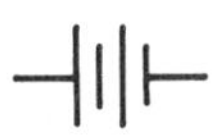

(b)

Figure 3–2 Battery. (a) A cell. (b) Circuit symbol.

Figure 3–3 Bulbs. (a) Neon type and circuit symbol. (b) Incandescent/illumination type and circuit symbol.

opposes a change in the existing current in a circuit. This opposition is called **inductance** (Fig. 3–5). Inductance is measured in henrys (H). Typical values of inductance in a coil are millihenries (mH) and microhenrys (μH). The reference letter for a coil is **L.**

A **transformer** is an electrical device with two coils wound closely together. Alternating-current (AC) energy is transferred from a primary winding to a secondary winding by **electromagnetic induction.** Transformers have the ability to step up (increase) or step down (decrease) an ac voltage or current. The frequency of the ac is not changed. The core of the transformer can affect its application and function.

Power circuits use iron-core transformers (Fig. 3–6a). Radio and television signals usually use air-core or adjustable-core transformers (Fig. 3–6b). Transformers are rated in henrys (H). The reference letter for a transformer is **T.**

Connectors fall into five basic categories: edge connectors for PC boards, cable

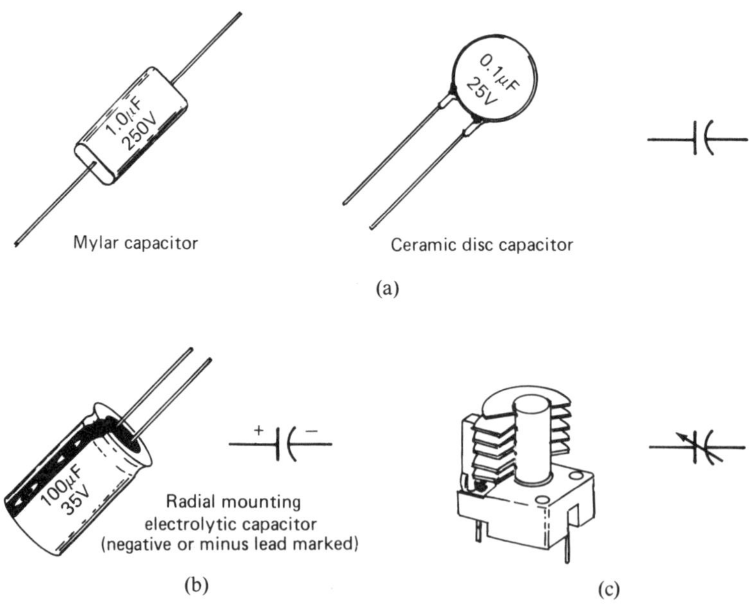

Figure 3–4 Capacitors. (a) Mylar and ceramic disc capacitors and symbol. (b) Electrolytic capacitor and symbol. (c) Variable capacitor and symbol.

Figure 3–5 Coils or inductors. (a) Fixed inductor and symbol. (b) Variable inductor and symbol.

connectors, ac power-line connectors, terminating connectors, and auto jacks (Fig. 3–7). There are numerous sizes and styles of connectors for each of these categories. Because of the numerous styles, there has been a problem developing a standard symbol representation. When designing a cable with connectors, consider whether it has a low- or high-power application, whether it will be subjected to low- or high-frequency signals, and whether it will be soldered, crimped, welded, or wire wrapped. The function of a connector is to provide a rapid connect or disconnect of the junction between two cables, a cable and PC board, or a cable and a chassis. The reference letter for jacks and connectors is **J.**

A **fuse** is a protective device that is sensitive to the amount of current being forced through it (Fig. 3–8a). When the amount of current is too great, an internal

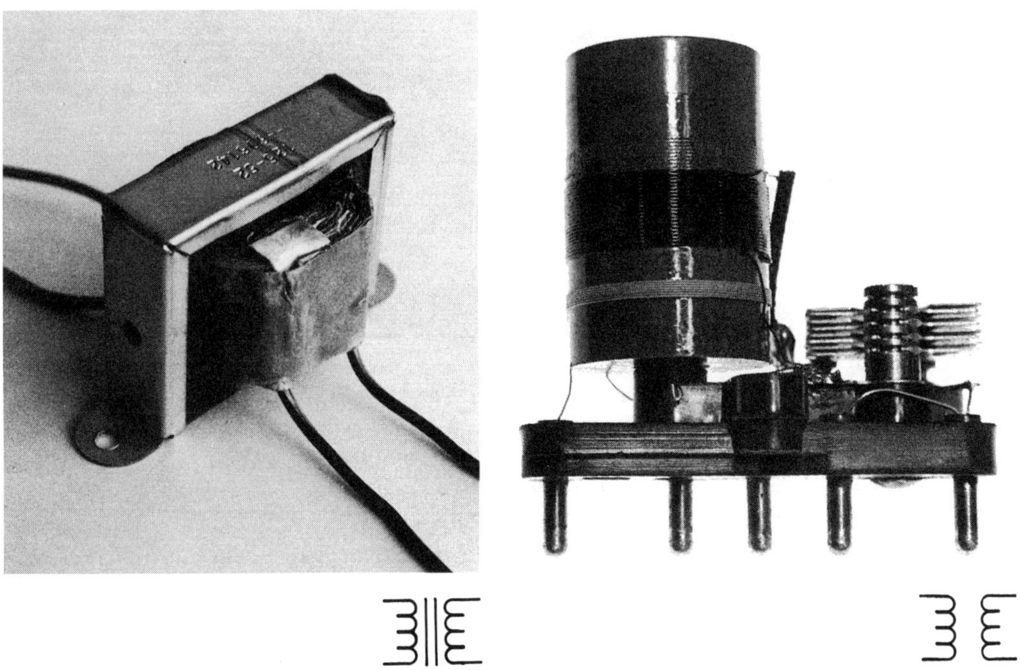

Figure 3–6 Transformers. (a) Iron-core transformer and symbol. (b) Air/adjustable-core transformer and symbol.

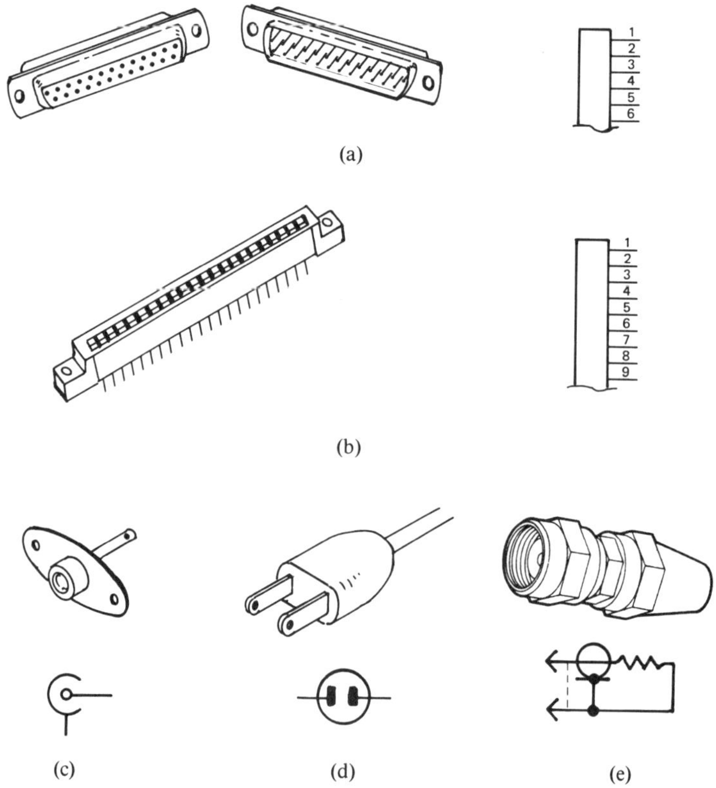

Figure 3-7 Examples of connectors and their circuit symbols. (a) Rack and panel. (b) PC board edge. (c) Phone or audio jack. (d) Plug. (e) Termination.

element, usually a short piece of wire, melts, causing the circuit to open. This stops the current flow just like an open switch.

An automatic fuse called a **circuit breaker** is shown in Fig. 3-7b. A circuit breaker opens when a current over its rated limit is experienced. The breaker's action may be caused by an overload coil or by a bimetal internal switch. Either type can be *reset* after a short time if the trouble causing the overload is corrected. The reference letter for fuse is **F.**

A **meter** is an electrical or electronic instrument that can measure current, voltage, resistance, and/or power (Fig. 3-9). There are two basic types of meters: analog, which uses a needle to indicate a measured value, and digital, which uses an LED or liquid crystal readout display.

A **microphone** is an electronic device that responds to sound waves by transforming changes in air pressure (movement of air) into electrical signals (Fig. 3-10). The Reference letters are **MIC.**

A **piezoelectric crystal** (Fig. 3-11) produces electricity when a varying pressure

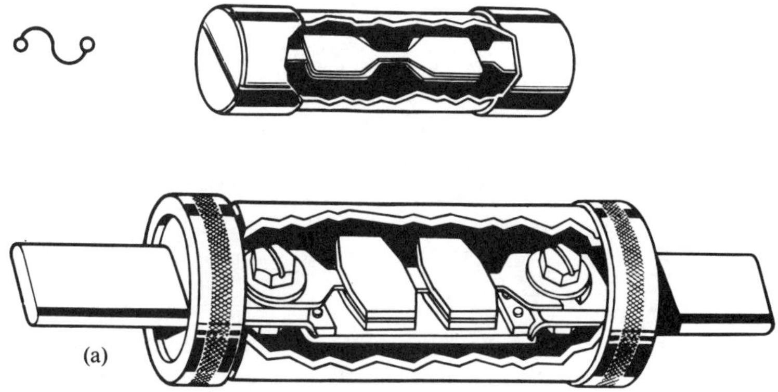

(a)

(b)

Figure 3–8 Fuses. (a) Fuse and symbol. (b) Circuit breaker and symbol.

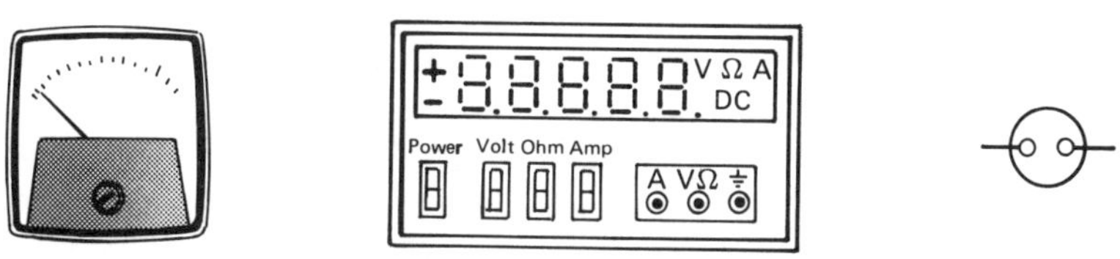

(a) (b) (c)

Figure 3–9 Meters. (a) Analog. (b) Digital. (c) Circuit symbol for both.

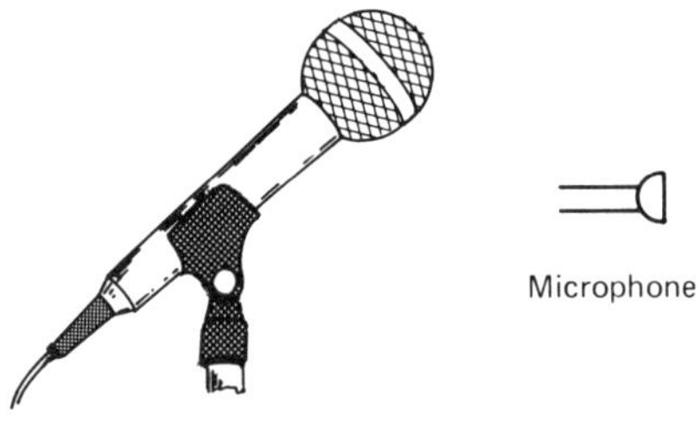

Figure 3–10 Microphone with its circuit symbol.

is applied to a crystallite material, such as natural quartz. It is used in radio and computer circuits to control the frequency of an oscillator. A crystal is designed to oscillate at one frequency. The crystal is used to generate a frequency that is used as a reference frequency in radio transmitters and as timing frequencies in computers. A crystal can also be used to change mechanical energy into electrical energy, as in a microphone. The reference letters for a crystal are **XTL.**

Resistors are components that have electrical properties of resistance. They are designed to limit or control the flow of current. The value of resistance is measured in ohms (Ω). Resistors are also rated by their ability to dissipate heat. The unit of measure for dissipating heat is the watt. The reference letter for a resistor is **R.**

Resistors are grouped in three categories: fixed, tapped or adjustable, and variable. The fixed resistor has only one resistance value and cannot be adjusted. Fixed resistors can be either carbon composition, wire wound, or film type (Fig. 3–12).

Carbon resistors are the most commonly used resistors (Fig. 3–13). They usually have narrow colored bands around one end to indicate their ohmic value and tolerance (degree of accuracy). Most resistors used today have a ± 5% tolerance. Some organizations require the use of only ± 1% resistors. Appendix A shows a standard color code chart. The color coding of resistors was covered in Section 1.2. Review that section if you have not learned the color code.

An adjustable or tapped resistor is a wire-wound resistor that has part of the resistance wire exposed to enable the amount of resistance in use to be adjusted as needed. Adjustment is made by sliding a movable contact across the exposed wire (Fig. 3–14).

A variable resistor can be continuously varied over the ohmic value of the device. For example, 100-ohm resistor can be adjusted anywhere from 0 to 100 ohms. Potentiometers, rheostats, and trimmers are all variable resistors (Fig. 3–15).

A **semiconductor** is a substance that at room temperature is neither a conductor or insulator. Silicon and germanium are two materials used in semiconductor devices.

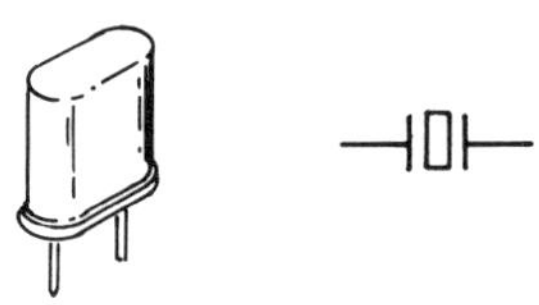

Figure 3–11 Piezoelectric crystal and its circuit symbol.

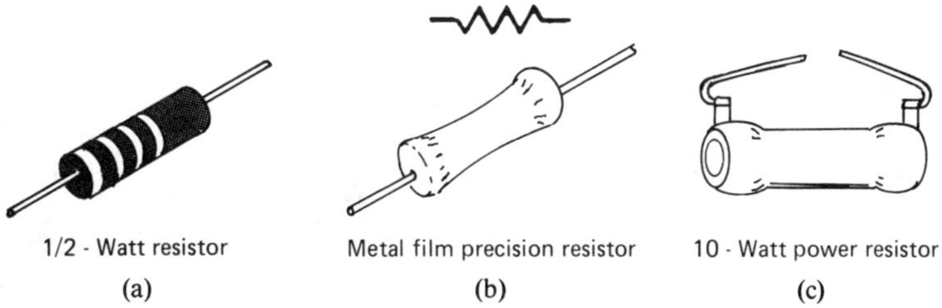

Figure 3–12 Fixed resistors and symbol. (a) Carbon resistor. (b) Film resistor. (c) Wire-wound resistor.

Pure semiconductor crystals are seldom used in electronics since they have few free electrons (allowing very little current flow). Therefore, small amounts of impurities are added (such as arsenic or boron) in a process called **doping.** The amount and type of impurities added determine the amount of current flow in the semiconductor device. The major function of a semiconductor is to control current. Semiconductor devices are economical and small in size, have low power consumption, and can perform many electronic and electrical operations. They are the foundation for consumer and commercial electronic systems. They are used in equipment such as radios, TVs, video games, calculators, computers, and the space industry.

A rectifier or diode is the simplest form of a semiconductor device. A diode conducts electrons more easily in one direction than in another (Fig. 3–16). It can act as a one-way switch to allow current to flow in only one direction. Its reference letters are **CR.**

Light-emitting diodes (LEDs) are used many different ways in electronics. They are used as numeric displays to indicate numbers in products such as calculators, clocks, digital meters, and other instruments. They are also used as pilot lights to indicated that equipment is energized. The reference letters are **LED** (Fig. 3–17).

A **transistor** is a three-element device with an emitter, base, and collector. Tran-

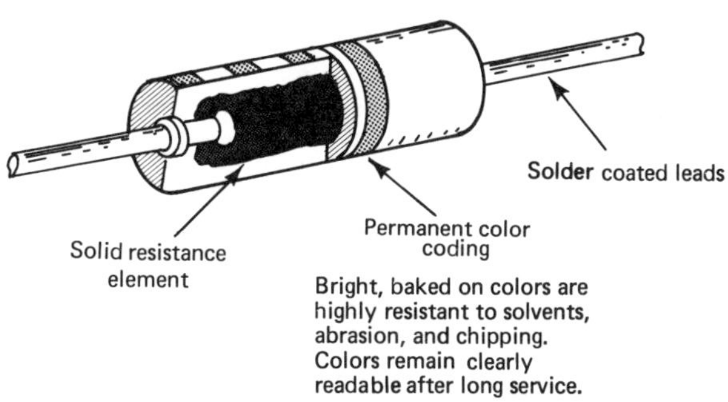

Figure 3–13 Carbon resistor.

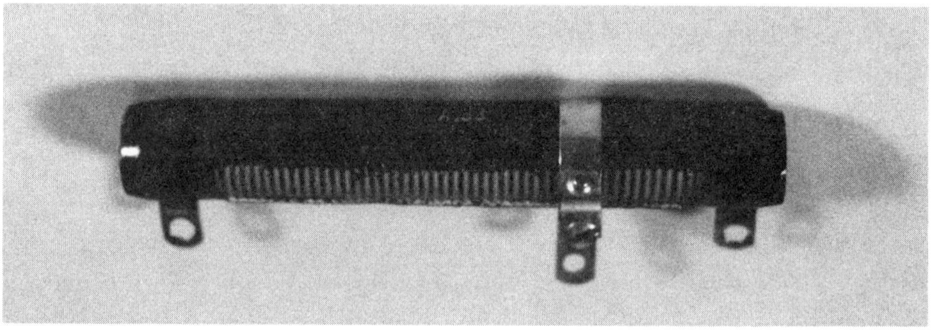

Figure 3–14 Adjustable or tapped resistors and symbol.

sistors come in many sizes and shapes, each with a specific function (Fig. 3–18). A transistor can be used as an amplifier, high-speed switch, oscillator, and a mixer. The letter designator for a transistor is **Q.**

Integrated circuits (ICs) have been one of the most significant technological developments in the electronic industry. Several major reasons are that they have reduced cost, increased reliability, and greatly reduced the physical size of electronic devices and equipment. Hundreds of transistors, diodes, resistors, capacitors, and other devices can be imprinted onto a single chip. ICs come in a variety of package styles. A few of the popular packages and their symbols are shown in Fig. 3–19. The reference letter for ICs is generally **U** for dual-in-line packages. Other devices, such as

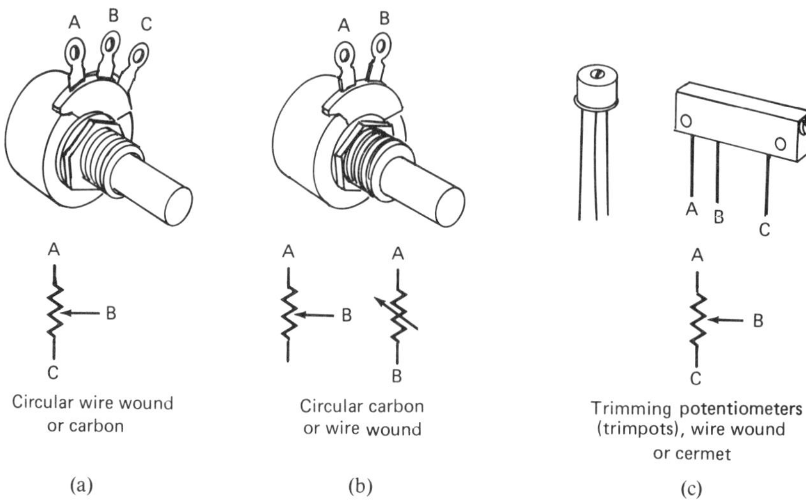

Figure 3–15 Variable resistors. (a) Pots and symbol. (b) Rheostat and symbol. (c) Trimmer and symbol.

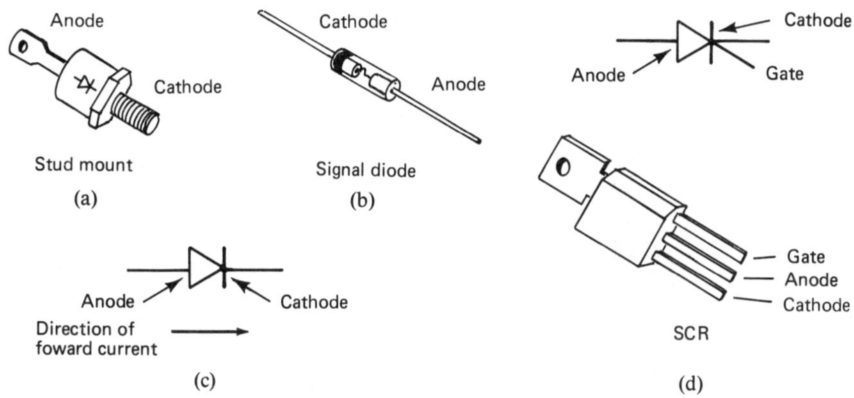

Figure 3–16 Diodes. (a) Power diodes. (b) Signal diode. (c) Schematic symbol. (d) SCR and schematic symbol.

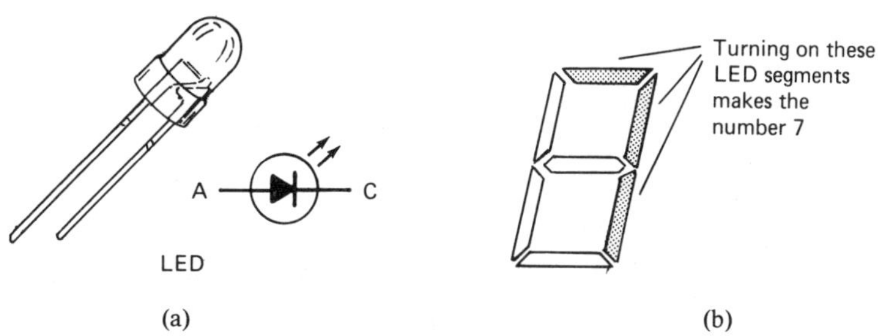

Figure 3–17 LEDs (light emitting diodes). (a) Single LED as a light indicator and symbol. (b) LED display.

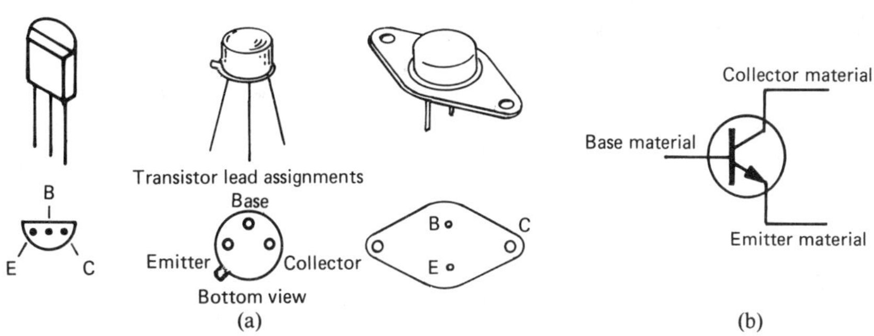

Figure 3–18 Transistors. (a) Common transistor case styles. (b) Schematic symbol, NPN transistor.

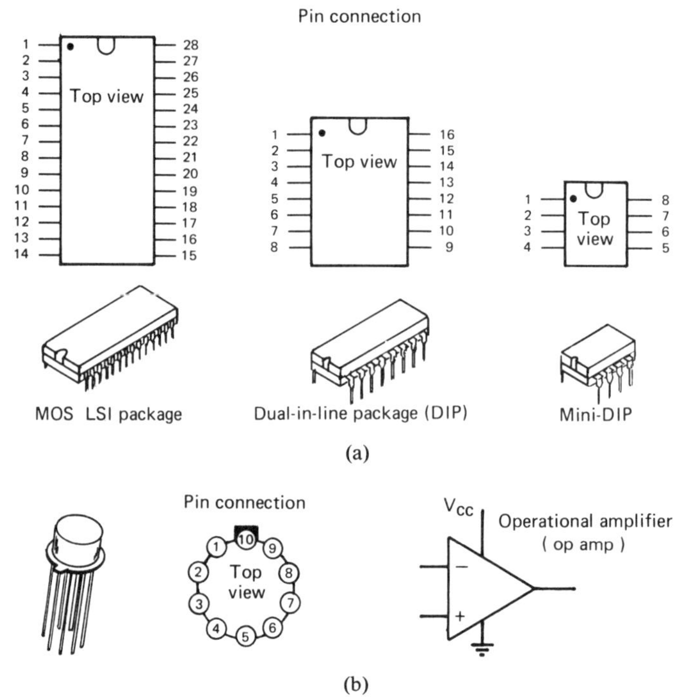

Figure 3–19 IC packages. (a) Three versions of dual-in-line packs with their schematic symbols. (b) TO-5 package with schematic symbol.

operational amplifiers, use the reference letters **AR.** See Appendix I for other common reference letters.

A **speaker** is an electrical component that has the ability to convert *electrical impulses* into sound (Fig. 3–20). The electrical impulses vibrate a paper cone in the speaker. This creates air movement in rhythm with the electrical impulses. The rhythmic air movements allow us to hear the sound that produced the electrical impulses that were fed into the speaker. The reference letters are **spkr.**

A **switch** is a mechanical or electrical device used to complete or break the path of current in a circuit or send it over a different path. There are many types of

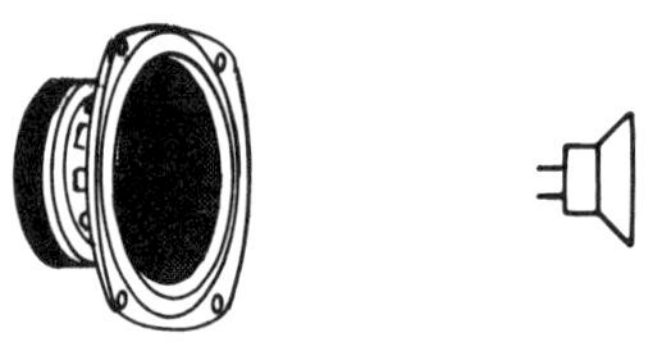

Figure 3–20 Speaker and symbol.

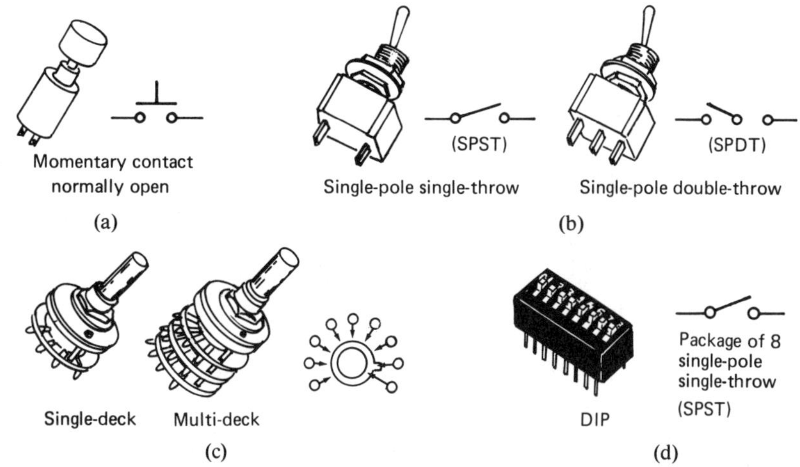

Figure 3–21 Switches. (a) Push button and symbol. (b) Toggle and symbol. (c) Rotary and symbol. (d) Dip switch and symbol.

switches that serve different purposes, such as push button, slide, toggle, micro, and wafer.

A switch can be momentary or on–off. A momentary switch is activated only when pressure is applied. A push-button switch can be either normal open (PBNO) or normal closed (PBNC) (Fig. 3–21a). A toggle switch is activated by an up–down action. They can be single-pole, single-throw (SPST); double-pole, single-throw (DPST); single-pole, double-throw (SPDT); double-pole, double-throw (DPDT); or larger, such as triple-pole, double-throw (3PDT). (Fig. 3–21b). The reference letters for a switch are **SW**.

A wafer switch is a rotary, multiposition switch with fixed terminals. The rotor arm is used to complete different current paths in a circuit by making contact with the terminals. Older-style televisions use a rotary switch for changing channels (Fig. 3–21c).

A **relay switch** is an electrical switch with one or more sets of contacts that are open or closed by electromagnetism (Fig. 3–22). As current flows through the coils,

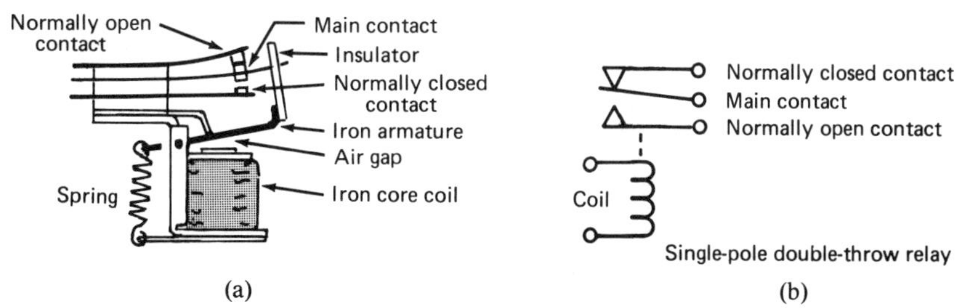

Figure 3–22 Relay switch. (a) Typical appearance. (b) Schematic symbol.

the magnetic field provides the mechanical action to the contacts to perform the switching function. Relays can control high-voltage circuits with low applied voltage. The reference letters for a relay switch are **RLY.**

A **terminal** is a device used as a point of connection for two or more conductors in an electrical circuit. Figure 3–23 shows four types of terminals. The reference letter for a terminal point is **E.**

A **vacuum tube** is a hermetically sealed glass or metal envelope in which conduction of electrons takes place through a vacuum or gas. Vacuum tubes were developed to control the flow of current. A vacuum tube is a vacuum enclosure containing a number of elements, such as a cathode, plate, and grid. Like transistors, tubes can amplify, rectify, or change the input signal (Fig. 3–24). The reference letter for vacuum tubes is **V.** The most often seen tube is the cathode-ray tube, which is used as the screen of the television set or for an oscilloscope.

A **cathode-ray tube** (CRT) is a visual display tube on a television or oscilloscope. A sweeping beam illuminates fluorescent atoms coated on the inner face of the tube to produce a visible picture or pattern. The reference letters for a cathode-ray Tube are **CRT** (Fig. 3–25).

3.3 STATIC AND ELECTRONIC DEVICES

Anyone working in the electronic industry should be aware of the cause and effect of electrostatic discharge (ESD). When components are unpacked, the action of moving the plastic packing material can cause an electrostatic charge. Electrostatic charges can also be generated on the body of a person by the rubbing of nylon materials together, walking across a carpet, or sliding across a plastic seat. The *static electricity* generated by any of these and many others things can be several hundreds of volts. This static electricity is only mildly annoying to humans—a snap as we reach for a door handle or a spark as we kiss our significant other. However, **static electricity is deadly to certain semiconductor devices** if it is allowed to discharge across them as they are being unpacked, removed from a circuit, or inserted into a circuit. This problem is of such severity in MOS-type semiconductor devices that it bears further discussion.

Static electricity is an electrical charge caused by friction. Static electricity is generated by material or body contact and separation, or by motion. These factors cause large numbers of electrons to be added to or deleted from a material, causing an unbalance of charges. When static charges become large, an electrostatic discharge takes place. Walking across a carpeted floor can generate as much as 12,000 volts. Discharge takes place when a charged body touches a neutral surface.

Friction is the most common cause of static electricity. A charged individual can easily discharge his or her supply of electricity to an uncharged or neutral body (PC board) and cause extensive damage. Components can also acquire electrical charges by this method. Charges can develop on devices from movement inside their shipping tubes. When a charged device is removed from the tube, a rapid discharge of electrical energy can damage the devices.

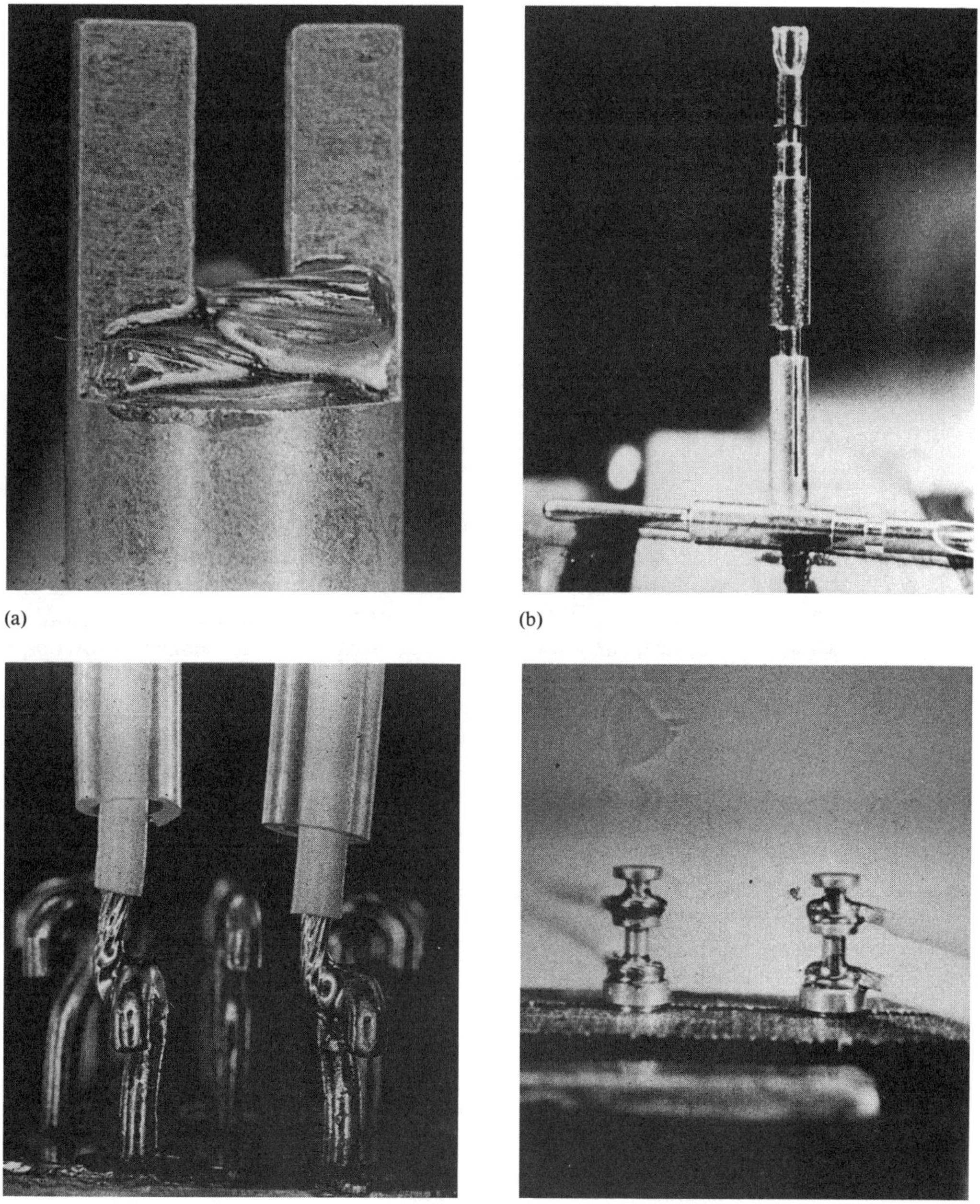

(a) (b) (c) (d)

Figure 3–23 Common terminals. (a) Bifurcated. (b) Cup. (c) Hook. (d) Turret.

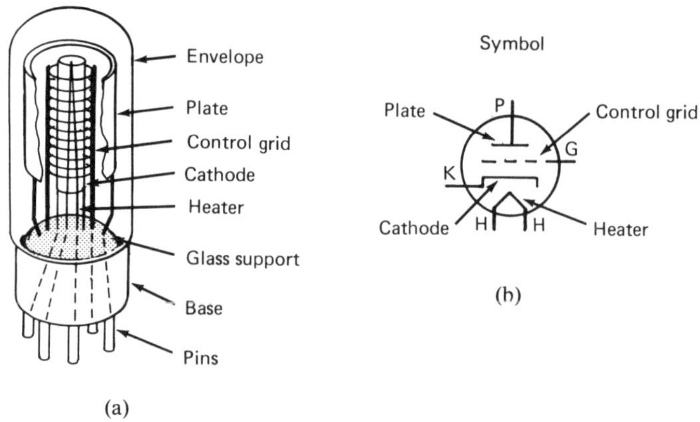

Figure 3–24 Vacuum tube. (a) Triode tube. (b) Schematic symbol.

A second type of static charge, **induction,** is developed on a conductor as a result of its proximity to a charged body. When a person handles a PC board or components wrapped in a plastic bubble-wrap material, a charge is induced onto the plastic, which then transfers a charge to the contents of the package. When the contents are removed from the wrap, the sudden discharge causes static damage to the components.

The third type of ESD is **static capacitance,** that is, static generated in relation to the distance between two conductive surfaces. The greater the distance is, the greater the voltage. When a PC board assembly is picked up and set back down (grounded) again, it is likely to be damaged by the large voltage generated when it was lifted from its original grounded position.

Metal-oxide semiconductors (MOSs) are extremely sensitive to static charges. They are categorized according to their reaction to static charges. Unprotected MOS devices, such as field-effect transistors (FETs) and large-scale integrated circuits

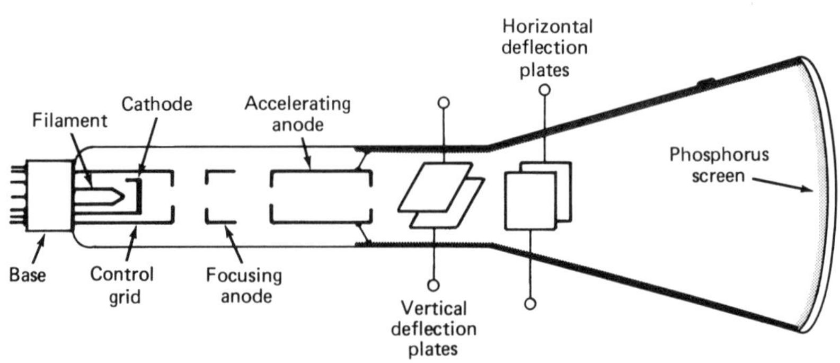

Figure 3–25 CRT and symbol.

(ICs), are in the *extremely sensitive category*. MOSs with protection networks, such as CMOS, Schottky diodes, and transistors, are in *the sensitive category*. The *least sensitive devices* include small-signal diodes, transistors, and piezoelectric crystals. Many other devices are sensitive to static charges; therefore, all components must be handled with care.

ESD control and prevention should be followed at all times when handling electronic components and circuitry. This is the responsibility of each individual employee. The following are some rules for good ESD control.

1. Handle all sensitive parts and components at a static-safe work station (Fig. 3-26).
   a. Use a common ground at each work station. The table mat, equipment, and so on, must be connected to the same grounding point.
   b. Use a conductive table mat and floor mat grounded through a 1-mehohm resistor. Use conductive heel straps or shoes with conductive soles when wrist straps cannot be used.
   c. All metal equipment, such as soldering irons, machinery, and electrical equipment, must be grounded.
2. Treat all electronic components and parts as static sensitive.
   a. Keep all parts in their original packages until needed.
   b. Before handling a PC board or devices, ground yourself by using a wrist strap grounded to a 1-megohm resistor (Fig. 3-27).
   c. Do not remove the protective shorting ring or band (grounding) from a device until it has been mounted and soldered on the PC board.

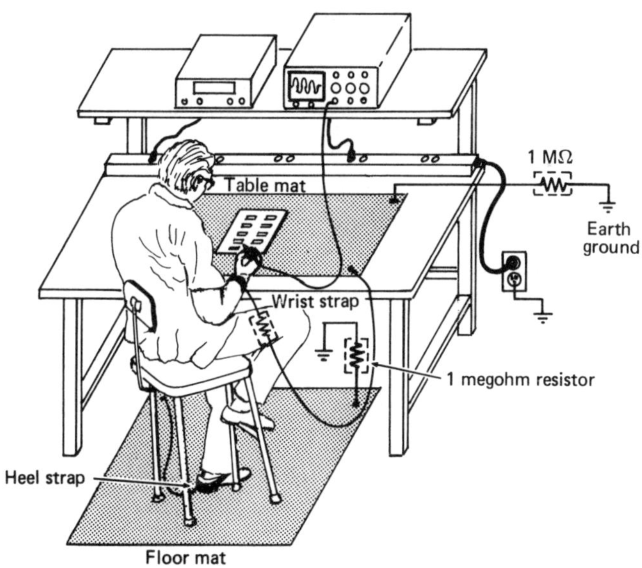

Figure 3–26 Static-free work station.

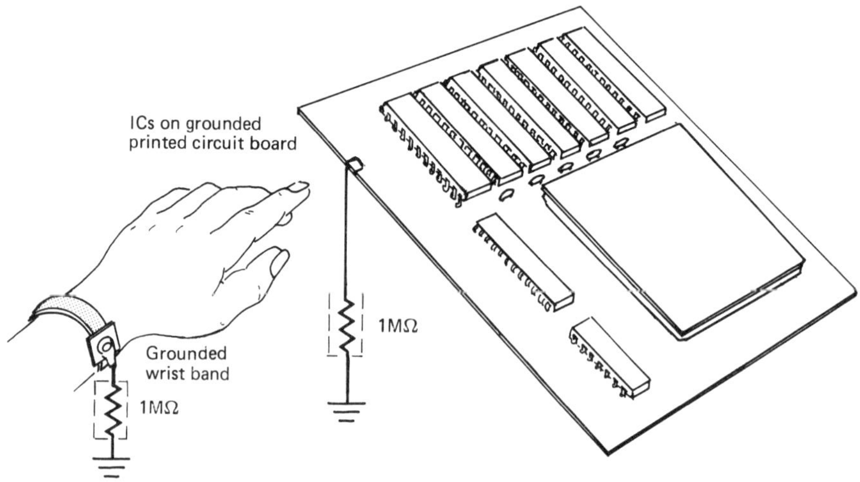

Figure 3–27 Grounded wrist strap.

d. Do not slide static-sensitive devices over surfaces.
e. Do not wear synthetic clothing. Smocks made of ESD-protective material are preferred, but short sleeves may be worn.
f. The work station must be clear of nonconductors, such as plastic products, cardboard, paper products, and food wrappers.

PARTS IDENTIFICATION TEST

Materials

1. Box of numbered parts, such as:

Battery	Turret terminal
Switch (SPST)	Meter
Transistor	Integrated circuit
Fuse	Illumination bulb
Edge connector	Speaker
Vacuum tube	Bifurcated terminal
Capacitor (dry)	Transformer
Diode (rectifier)	Capacitor (electrolytic)
Switch (toggle)	Inductor (air-core) coil
Antenna	Microphone
Potentiometer	Switch (push button)
Switch (wafer)	Resistor (fixed)

Neon bulb	Piezoelectric crystal
Panel connector	Light-emitting diode

2. Electronics Dictionary

Procedure

Identify each numbered part provided by your instructor. Give the identification number, name, circuit symbol, reference letter, and description of its function, as appropriate.

Chapter 4

Soft Soldering

4.1 INTRODUCTION

Soldering is the process of joining two or more metals at a temperature below their melting points. Soft solder is an alloy composed of tin and lead, which is manufactured in a number of percentage combinations. For example, when the proportions are equal, it is known as 50/50 solder. The most commonly used combination in the electronic industry is 60% tin and 40% lead, expressed as 60/40 solder. What makes solder so useful is a quirk of nature. Tin has a melting point of approximately 450°F, and lead melts at 621°F. When added together, the melting temperature is less than that of either. In a 60/40 combination, solder melts at 361°F. Different combinations of tin and lead have a different melting point. This can be seen in the tin–lead fusion chart in Fig. 4–1.

Commercial soft solders sometimes contain various amounts of antimony, bismuth, cadmium, or silver, which are added to change the physical properties of the tin–lead combination. However, many solders contain small amounts of these elements only as impurities. Solder is available in many forms, such as the following:

- Cake or ingots: Rectangular or circular in shape weighing 3.5 and 10 pounds. Used for large-volume applications requiring the addition of intermediate quantities during production.

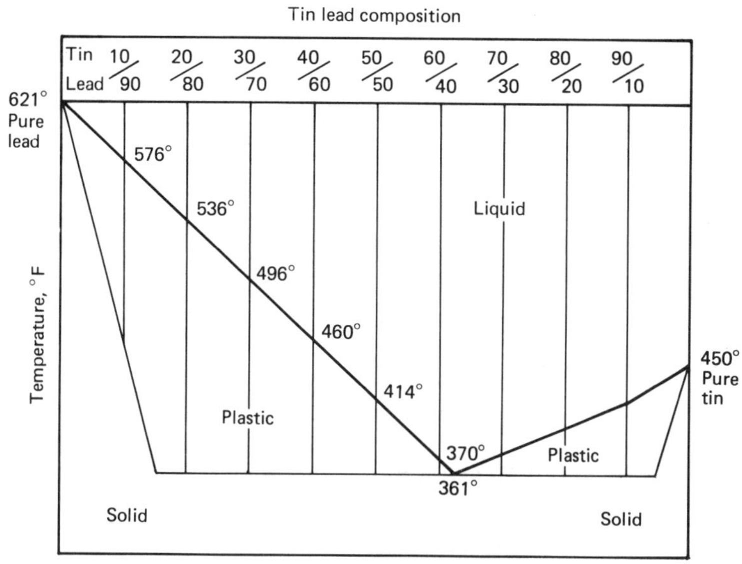

Figure 4–1 Solder fusion chart.

- Bar solder: Available in weights from $\frac{1}{2}$ to 10 pounds and is generally used for wave solder machines.
- Paste solder: Available as a mixture of solder powder and flux in a paste form.
- Foil, sheet, and ribbons solders: Supplied in various thicknesses and widths.
- Wire solder: Available in diameters from 0.005 to 0.250 inch on spools weighing 1 to 5 pounds.
- Flux core wire solder: Cored with organic, inorganic, or rosin fluxes. Available in diameters of 0.10 to 0.250 inch on spools weighing 1 to 20 pounds.
- Acid Fluxes: **Never use for electronic work. They are corrosive and will cause defective solder connections.**

Solder is used to bond wires to copper traces (conductors), to bond wires together, and to bond wires to terminals or fasteners. The size of solder used depends on what is being soldered. For example, to solder a large wire to a large terminal, use a large-diameter wire solder. To solder a component to a printed circuit board, use a small-diameter wire solder.

When sufficient heat is applied to solder, it changes from a solid to a plastic state and then to a liquid. Soldering takes place at the liquid state. Solder solidifies in the reverse manner. That is, it does not solidify instantly but goes from liquid to plastic to solid when heat is removed (Fig. 4–1). If the joint being soldered is moved during the cooling period, the solder becomes coarse grained and produces a poor connection called a **fractured joint.** If the heat is removed before the solder completely liquefies, the connection is commonly referred to as a **cold joint.** On the other hand, solder kept in the liquid state too long causes the solder to turn granular and gray in color. This is called a **hot joint.** These solder connections are unacceptable in electrical circuits and could cause circuit failures. (Fig. 4–2).

When heat is applied to a solder connection and solder is applied, there is a chemical and a *metallurgical* or *metal solvent action* between the solder and the metals being joined. The heated solder flows around the joint and mixes with the molecules on the surface of the metals being joined (a small amount of copper melts and mixes with the solder). The properties of a soldered joint, therefore, are different from the original solder due to the metallurgical action in the soldering process. The solder is partly converted to a new and different alloy. When the connection cools, the two metals fuse together to become a continuous metal surface. Fusion of the two metals must take place to form a good electrical connection (Fig. 4–3).

Before soldering takes place, the joint must be clean. The parts to be soldered must be cleaned of oil and dirt. This cleaning may be accomplished with a solvent. Isopropyl alcohol is one of the most commonly used solvents. The wires and terminals used in electrical circuits are usually made of copper. When exposed to air, a nonmetallic film called oxide forms on the surface. Oxides form an insulating barrier that prevents metals from touching each other. As long as this nonmetallic oxide barrier exists on the metal surface, the metals themselves cannot make metal-to-metal contact. These oxides must be removed before solder will effectively bond wire and terminals together. They must be cleaned off mechanically and/or chemically. An abrasive eraser such as a typewriter eraser is one kind of mechanical cleaner. A component lead cleaner made from plated copper braid is another. These mechanical cleaners are shown in Fig. 4–4.

The most important cleaner is called **flux.** Whenever parts are soldered to-

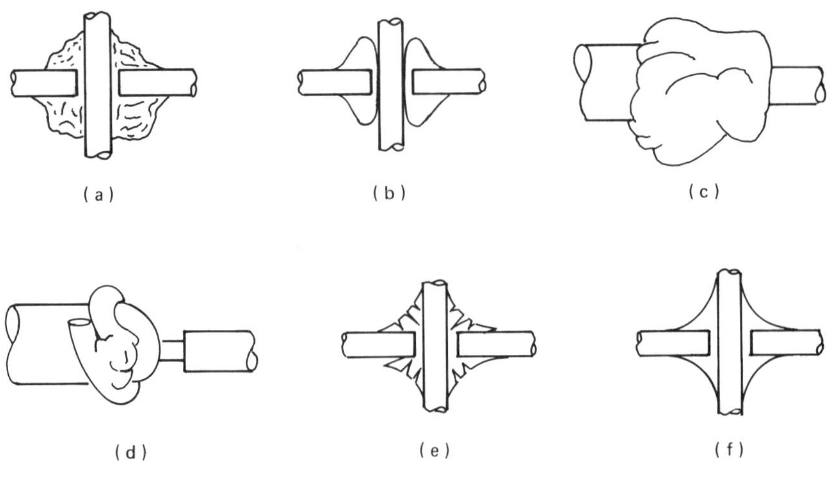

Figure 4–2 Solder joints. (a) Hot or overheated. (b) Cold joint. (c) Excessive solder. (d) Too little solder. (e) Fractured joint. (f) Acceptable joint.

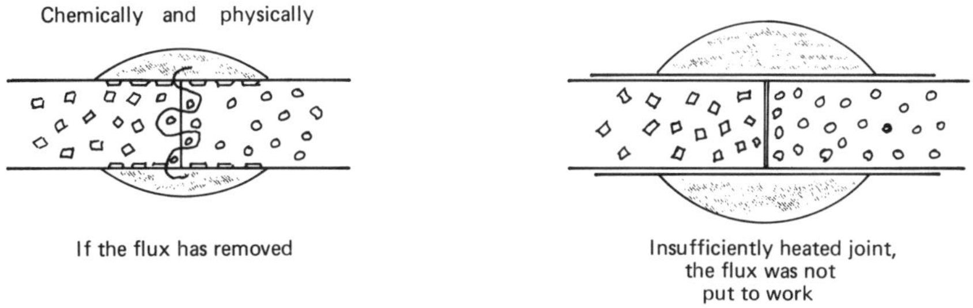

Figure 4–3 Fusion of two metals.

gether, flux is used to help remove the nonmetallic oxides and allow the cleaned metals to make mutual metallic contact. For convenience, the flux is usually inside the solder. The flux (which is made from tree rosin) melts at a lower temperature than solder. When the soldering iron touches the joint to be soldered, the joints get hot. When the solder is applied, the flux melts, cleaning the area of oxides, and then the solder melts and flows freely over the clean surface (Fig. 4–5c).

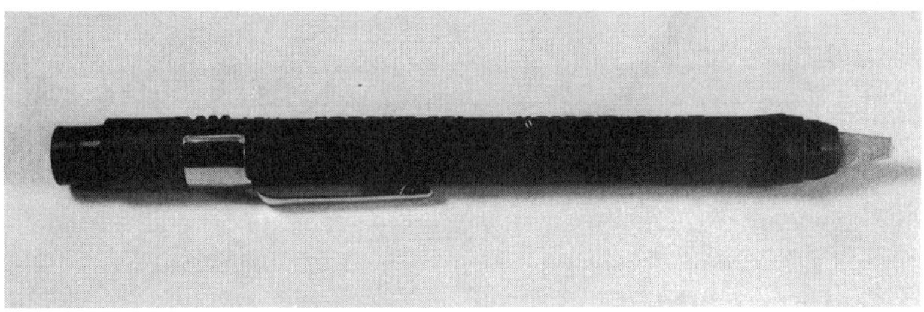

(a)

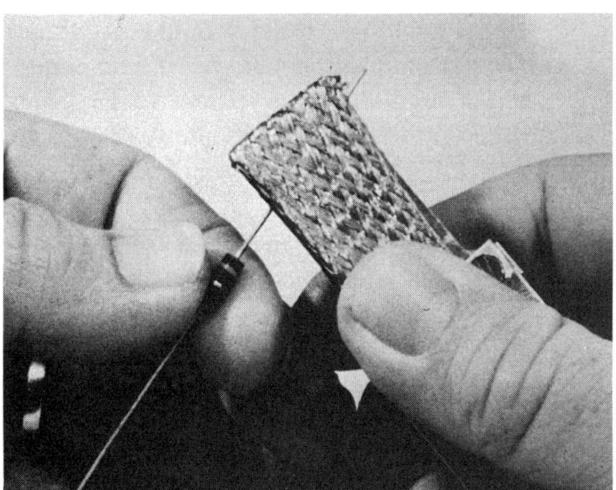

(b)

Figure 4–4 Mechanical cleaners. (a) An eraser. (b) Component lead cleaner.

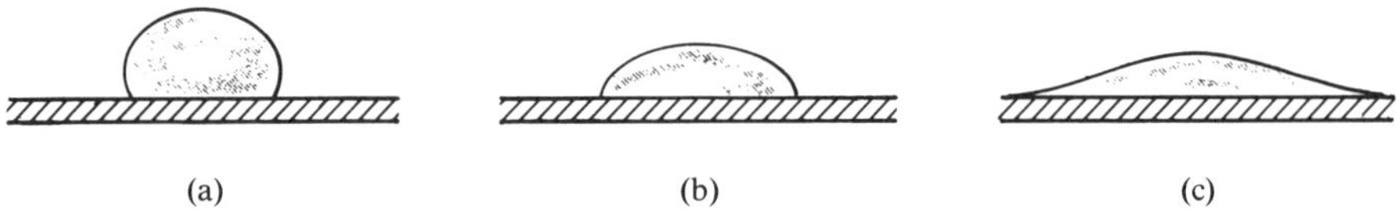

Figure 4–5 Wetting action of joints. (a) Nonwetting. (b) Partial wetting. (c) Total wetting.

On a thoroughly clean sheet of copper, a marked difference can be noticed in the reaction of the solder. On the unclean sheet of copper, solder balls up and does not flow as smoothly as it should. A cross section of a soldered joint shows large areas where the two metals have not blended together. The mixing of the molecules, or the *wetting action,* has not taken place (Fig. 4–5b).

In addition to being included in the solder, flux is available in solid, paste, or liquid form. In this way, additional flux is available if it is needed. After the solder has cooled, the flux and residue can be cleaned away with isopropyl alcohol or other flux cleaner. Flux will only remove metal oxides. It will not remove such materials as paint, shellac, sulfides, excessive dirt, or other inert matter from copper surfaces.

There are many types of soldering fluxes. The process of selecting the appropriate type can be complicated, depending on the size of the conductors, product specification, and design requirements. The size of the flux core can affect the outcome of the soldered joint (Appendix B). Generally, the smaller the point to be soldered, the smaller the diameter of the flux core. Some solders have a multicore rather than a single core. Figure 4–6 shows a chart of different-cored solder.

Regardless of the type of core selected, **only rosin fluxes should be used for all electrical connections.** Appendix C lists three common fluxes and their properties. The use of water-soluble and acid-type fluxes is prohibited for hand soldering. Military specifications require the use of RMA (active rosin noncorrosive flux) fluxes because they are nonactive at room temperature, and if they should become entrapped beneath components (called rosin joints), they will not hinder the reliability of the product. Commercial specifications are not as rigid; therefore, other types of rosin fluxes may be used. Other military specifications for fluxes and solvent cleaners are as follows:

- Liquid flux used with flux-cored solder shall be chemically compatible with the solder core flux and with the materials with which it will come in contact.
- The use of liquid rosin flux is restricted to the *wicking off* procedure for rework

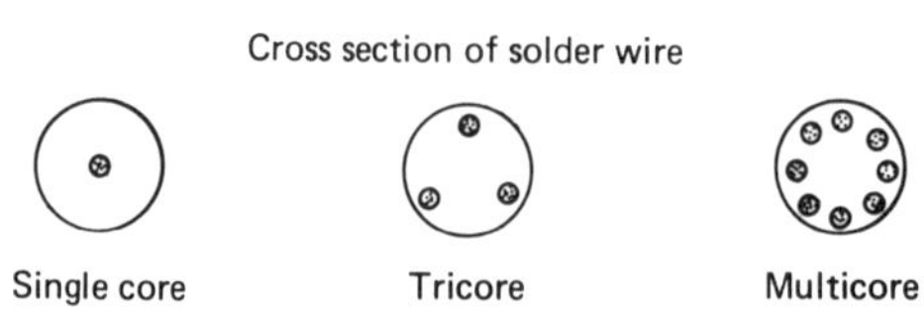

Figure 4–6 Flux cores.

of soldered connections, tinning operations, reflow soldering, and automatic machine soldering.

- Flux or solvents shall not be used in any manner that will carry to or deposit residue on contact surfaces, such as those in switches, potentiometers, or connectors.
- Acceptable solvent cleaners are ethyl alcohol, ACS grade 99.5% or 95% by volume; isopropyl alcohol, best commercial, 99% pure; and trichlorothifluoroethane, clear, 99.8% pure.
- **Sonic or ultrasonic vibration shall not be used as a method for cleaning electrical or electronic parts or assemblies unless it has been demonstrated that the reliability of the parts or assemblies will not be degraded by the process to be used.**

Regardless of the solder, flux, and solvent used in electronic circuit construction, in the final analysis the most important factor is product quality and reliability. Therefore, good soldering skills are a must. This chapter will teach you how to make reliable solder connections and build your soldering skills.

4.2 SELECTING THE PROPER SOLDERING IRON

Soldering irons come in a variety of sizes, shapes, and wattage, but consist basically of three main elements: a resistance heating unit; the heater block, which acts as a heat reservoir; and the tip, for transferring heat to the work. In the standard iron, the input voltage is fixed and constant, so the resulting tip temperature depends on the capacity of the heating unit and the size of the tip block (Fig. 4-7).

With more elaborate irons, the operator can increase or decrease the voltage applied to the heater and vary the tip temperature. Another variation is an iron that contains a temperature sensor built within the block, allowing the operator to set the desired temperature. The power to the resistance heater is automatically turned off and on to maintain the tip at the desired temperature.

Frequently, soldering irons are selected on the basis of input wattage or tip *idling* temperature, neither of which is a valid criterion for matching the right iron to the job. What is important is the tip *operating temperature* while it is actually heating the connection to be soldered. The ideal tip working temperature, or *tip thermal working zone,* is defined as a temperature range between approximately 600° and

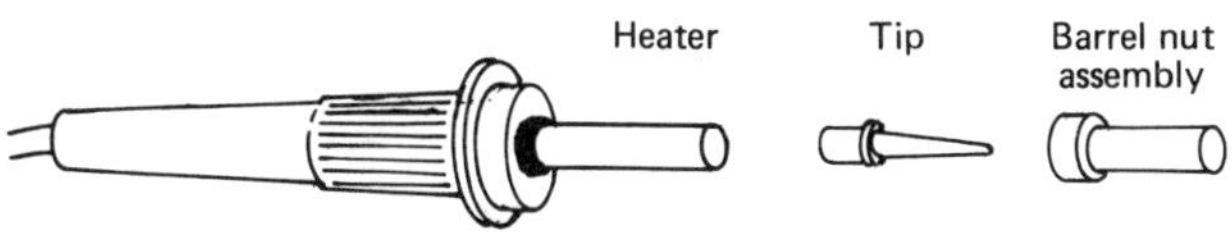

Replace tip screw barrel finger tight. The tip should be just tight enough to secure maximum heat transfer between the heating element and the tip.

Figure 4–7 Basic elements of a soldering iron.

700°F for most electronic assembly. As the operator begins to solder and makes successive soldered connections, the tip will reach an equilibrium (balanced) temperature within the thermal working zone if the soldering iron and tip are properly matched to the job.

4.3 CARE OF THE SOLDERING IRON

- Soldering irons should be turned off or unplugged during prolonged periods to prevent the buildup of oxides and reduce the chance of fire.
- Do not tap the iron on the bench or holder; clean only on a wet sponge.
- The soldering iron is a precision tool; do not use it as a crowbar or screwdriver.
- When replacing the tip, make sure it is fully inserted; otherwise you could damage the element.
- Always use the proper screwdriver if one is required to change tips or fittings.
- If the soldering iron has a collar sleeve, use the proper pliers. Do not use too much pressure; it could deform the case.
- The soldering iron must be stored safely between soldering cycles, preventing any injury to the operator.
- The soldering iron should rest in the holder so that the tip is in free air to minimize heat loss due to heat sinking.
- The inside of the holder should be clear of projecting burrs that would damage the tip.

4.4 SOLDERING TIP CARE AND MAINTENANCE

A high-quality soldering iron tip is made of copper for good conduction and, as with most soldering irons, is iron coated to prevent pitting. The following are some basic rules for care of the soldering iron tip:

- *Tin* the soldering iron by applying solder and flux to the clean tip the moment it is hot enough to liquefy the solder (Fig. 4–8).
- Never let the iron lose its tinning. The tinning action coats the tip with solder and protects it from oxidation.
- Do not overwipe the tip; use quick light strokes.
- Do not burn the sponge; keep it wet with tap water.
- After wiping the tip, apply fresh solder to the sides.
- Never wipe the tip before putting the iron in the holder; solder protects the tip during storage.

(a) (b)

Figure 4–8 Tinning. (a) Apply solder to the tip of the soldering iron. (b) Wipe tip on sponge or damp cloth.

- A poorly tinned tip will take too long to make each connection and overheat the connection.

4.5 PREPARATION FOR SOLDERING

4.5.1 Preparation of Conductors

1. Tin the tip of the iron prior to and during the soldering process.
2. Clean all surfaces of grease, dirt, oxide, or other foreign materials prior to soldering.
3. All wires, component leads, and terminals that do not have an easily solderable coating shall be tinned before soldering.
4. All leads and wires shall be formed and trimmed to proper length before soldering.
5. All trimmed leads shall be protected by solder tinning or other protective coating.
6. Stripping machines or hand tools used to remove conductor insulation shall be of the correct size and in correct adjustment and/or calibration.
7. After insulation removal, the remaining conductor insulation shall not exhibit any damage such as nicks, cuts, crushing, or charring. Conductors with damaged insulation shall not be used. Slight discoloration from thermal stripping is acceptable.
8. The lay of wire strands shall be restored as nearly as possible to the original lay if disturbed. Contact with bare fingers shall be avoided; however, if contact is made, the wire shall be cleaned with an approved solvent prior to further processing.
9. The portion of stranded or solid wires and part leads that will eventually be-

come a part of the finished solder connection shall be tinned with hot tin-lead solder and cleaned prior to attachment.

10. Additional flux may be used. The flux shall be applied so that it does not flow under the conductor insulation to prevent *solder wicking*.
11. Flux shall be removed using approved cleaning solvents. Position the wire so that gravity helps keep the solvent from flowing under the insulation.
12. Use an antiwicking tool to prevent solder from running under the insulation.

4.5.2 Preparation of Terminals and Solder Cups

1. Terminals and solder cups shall be examined and cleaned immediately prior to attachment of conductors.
2. Terminals and solder cups shall not be modified to accommodate improper conductor sizes.

4.6 SOLDERING TECHNIQUE

4.6.1 Placement of the Iron

1. Choose the appropriate tip size, shape, and heat range (Fig. 4–9).
2. Place the iron to allow for maximum heat transfer to both base metals (Fig. 4–10a).
3. Place the tip of the iron opposite in relation to the wire insulation (Fig. 4–10.b)
4. Apply the solder to the upper portion of a junction so that the heat of the work surfaces (wire, terminals, and the like) and not the heat of the iron melts the solder. **Never** apply the solder directly to the iron (Fig. 4–10c).

4.6.2 Sequence of Soldering

1. Apply the iron (heat terminal).
2. Apply the solder (when the terminal reaches the melting temperature of solder).
3. Remove the solder (as soon as the proper amount of solder has been applied).

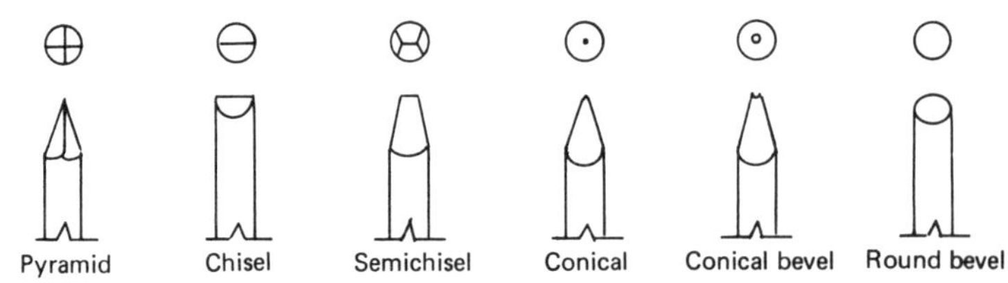

Figure 4–9 Common soldering iron tip shapes.

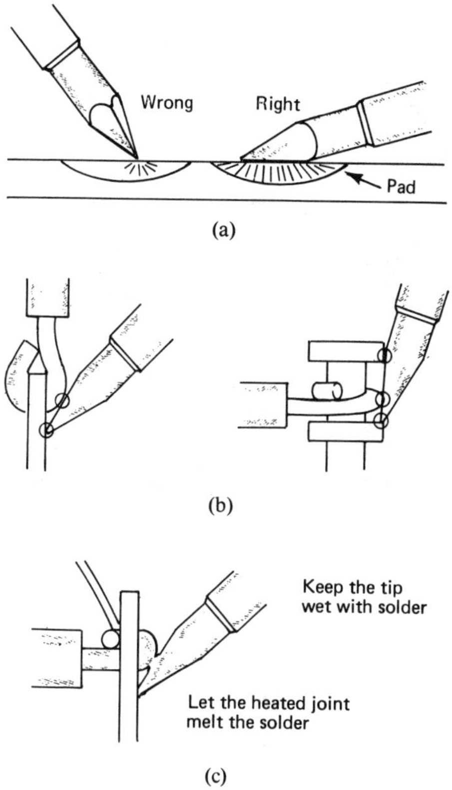

Figure 4–10 Placement of iron. (a) Note angle of the iron for best heat transfer. (b) Placement of iron in relation to wire insulation. (c) Placement of iron in relation to the solder.

4. Remove the iron.
5. Cool the joint. [Do not disturb (move) the joint until cooled. Movement could result in a fractured joint.]
6. Do not blow on the joint.

4.7 TINNING A WIRE

Tinning is the wetting of a surface with a thin, uniform layer of solder. Tinning is done to improve solderability, protect parts from oxidation, and prevent wires from birdcaging (spreading). Two methods can be use for tinning:

1. Soldering iron
2. Dipping in a solder pot

Before tinning a wire, the insulation must be removed with wire strippers. When using mechanical wire strippers, place the wire between the jaws and squeeze the handle (Fig. 4–11).

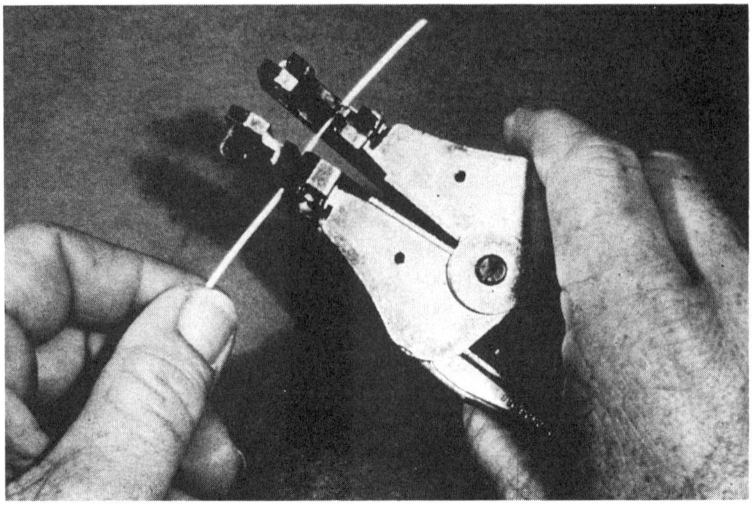

Figure 4–11 Using dual blade wire strippers.

To remove the insulation with thermal strippers, place the insulated wire between the electrodes (Fig. 4–12) and squeeze gently to close the electrodes. Turn the wire slightly while removing the insulation to even the cut. NASA and military standards do not recommend using mechanical strippers, because mechanical strippers may nick wires, causing the wire to break.

The following describes the proper sequence for tinning a wire:

- Prepare the tip for soldering (see Section 4.2).
- Restore wire strands to the original lay as much as possible.
- Use an antiwicking tool to support the wire and prevent solder from running (*wicking*) under the insulation (Fig. 4–13a).
- A *heat sink* may be used in place of the antiwicking tool (Fig. 4–13b).
- Before tinning the wire, coat the wire with liquid flux.
- Melt a drop of solder, called a *solder pool,* on the cleaned tip (Fig. 4–14).
- Add solder to the wire as you pull the iron slowly toward the end of the stripped wire (Fig. 4–15).
- Examine the wire to see that each strand is visible under the solder and that solder did not wick under the insulation.

4.8 SOLDERING WIRES AND CONDUCTORS TO TERMINALS

Terminals serve as tie points for wire junctions. Terminals may be mounted on insulating material such as Bakelite and porcelain or on PC boards. **Turret** and **bifurcated**

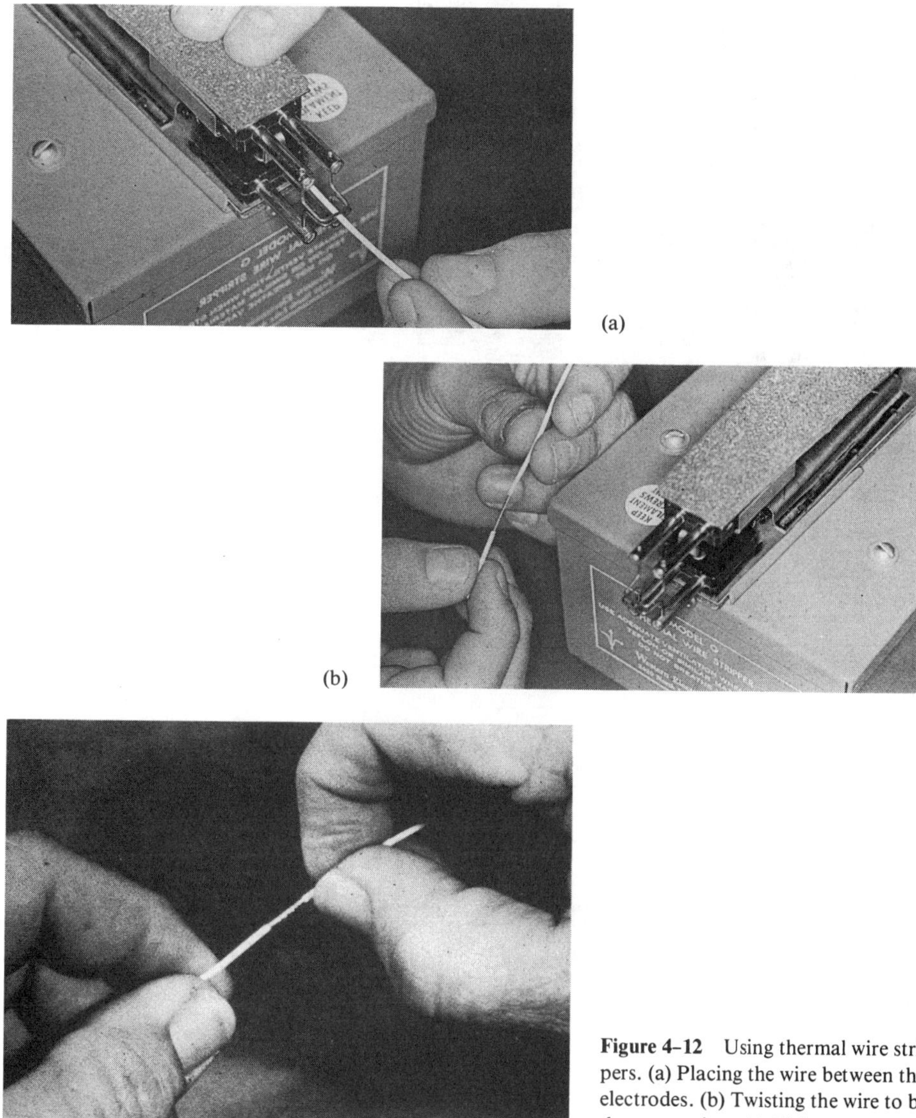

(a)

(b)

(c)

Figure 4–12 Using thermal wire strippers. (a) Placing the wire between the electrodes. (b) Twisting the wire to break the connection. (c) Sliding the stripped insulation off the wire.

terminals are ideal as tie points when one or two connections are to be made. But when many wires come to a PC board or some other point in a system, multiterminals (terminal strips) are used. Once you have learned how to solder a turret terminal, soldering to other terminals will be easier because the soldering process is similar for all terminals. The only exception is cup terminals. They require a different process,

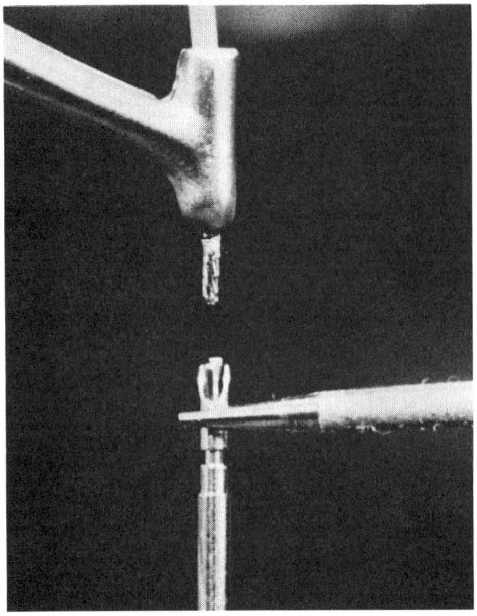

(a)

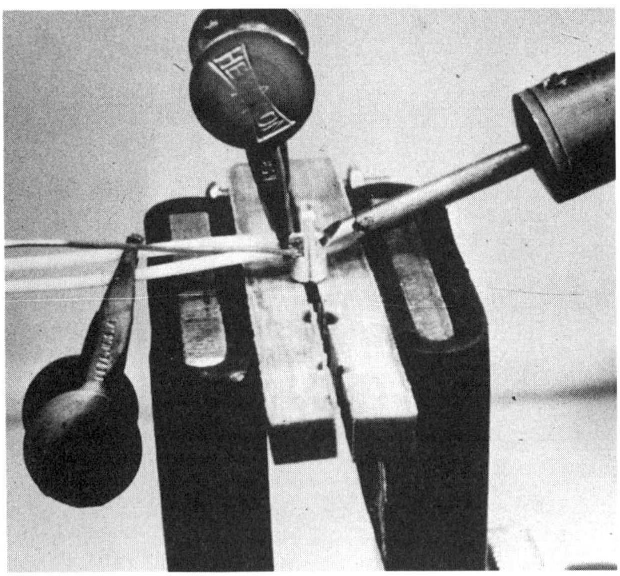

(b)

Figure 4–13 Protecting wire insulation. (a) Antiwicking tool. (b) Heat sink.

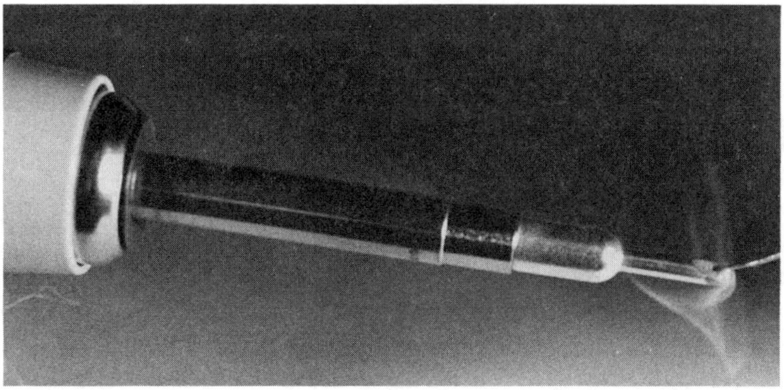

Figure 4–14 Applying a pool of solder.

which will be explained later in this chapter. Figure 4–16 shows several type of standard terminals.

Some general rules for attaching conductors to terminals are the following:

- For multiple wires routed to form a common cable trunk to equally spaced terminals (breakout cable), the length of the conductor ends should be uniform to prevent concentrations of stress on any one wire.
- The insulation shall not be embedded in the solder joint. The insulation clearance (*strip gap*) should not exceed the wire diameter, including the insulation (Fig. 4–17).

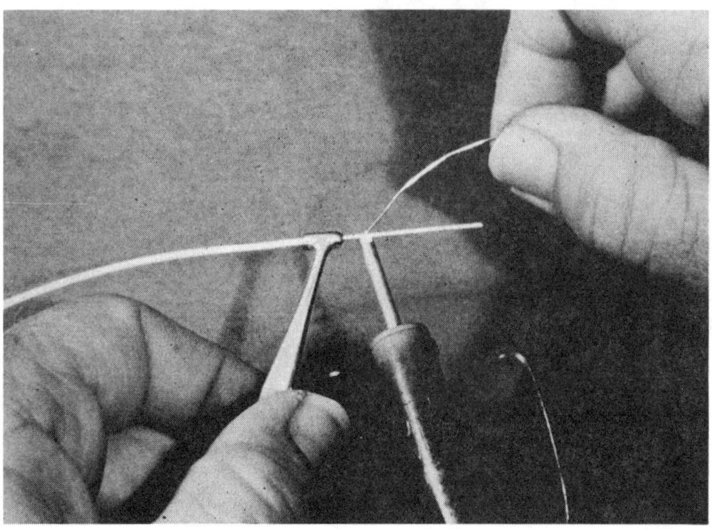

Figure 4–15 Tinning a wire.

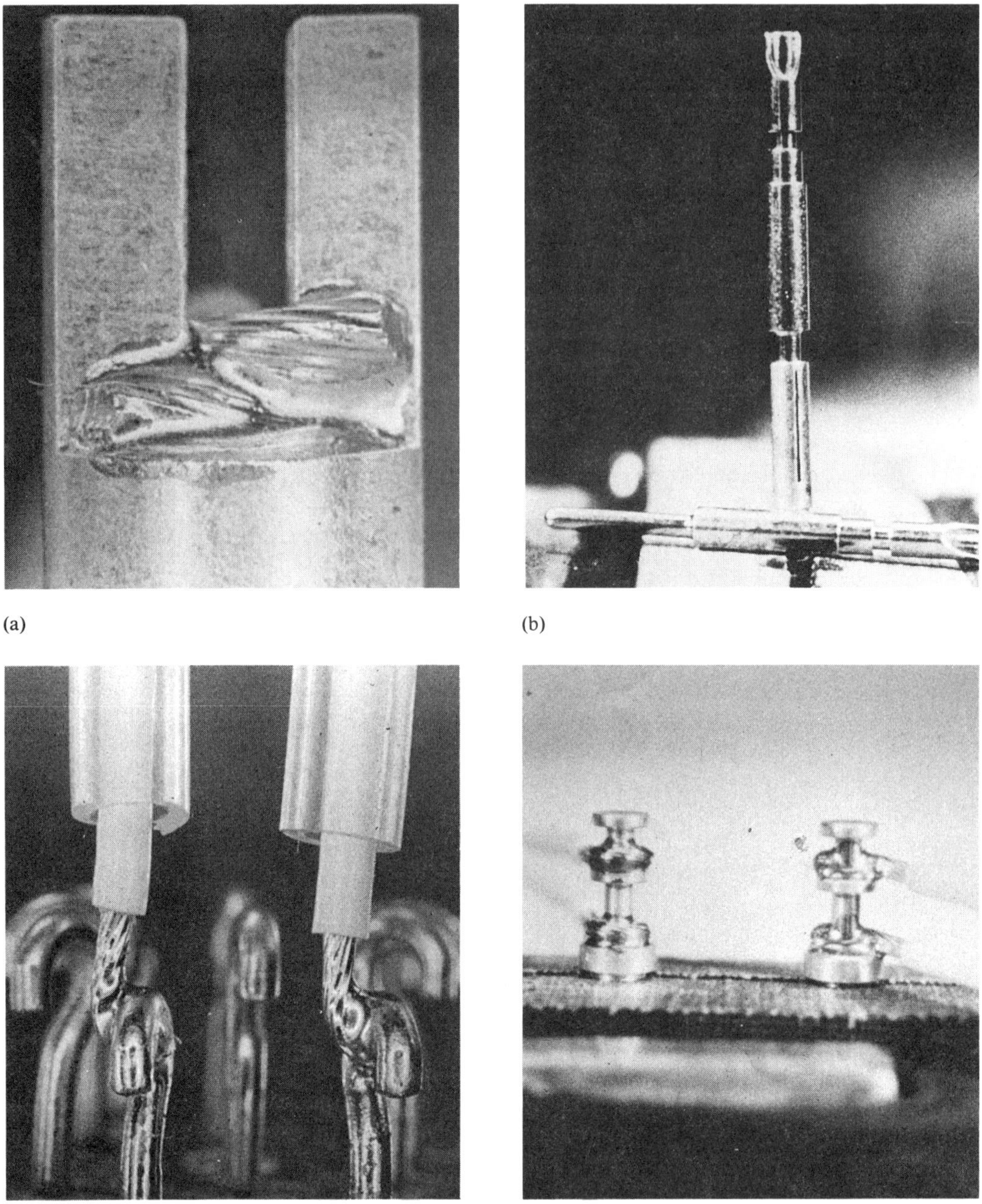

(a) (b) (c) (d)

Figure 4-16 Common terminals. (a) Bifurcated. (b) Cup. (c) Hook. (d) Turret.

- Use antiwicking tools (heat sinks) to prevent solder from running under the insulation.
- If there are multiple parallel entries of wires to a terminal, insulation clearances need not be equal.
- Each conductor terminating at a solder connection should have provisions for *stress relief* to minimize stresses to the connection during thermal or mechanical variations.
- Terminals should be of proper size to accommodate the conductors.
- Terminals and conductors should not be modified to accommodate improper sizes.

4.8.1 Turret Terminals

When soldering to a turret terminal, the areas to be soldered must be cleaned of oils, greases, and oxides. Usually, the terminals are plated with a metal that does not oxide as fast as copper so that the surface is nearly free of oxides. However, a mechanical or chemical cleaner should be used to prepare the surface for soldering. When solder-

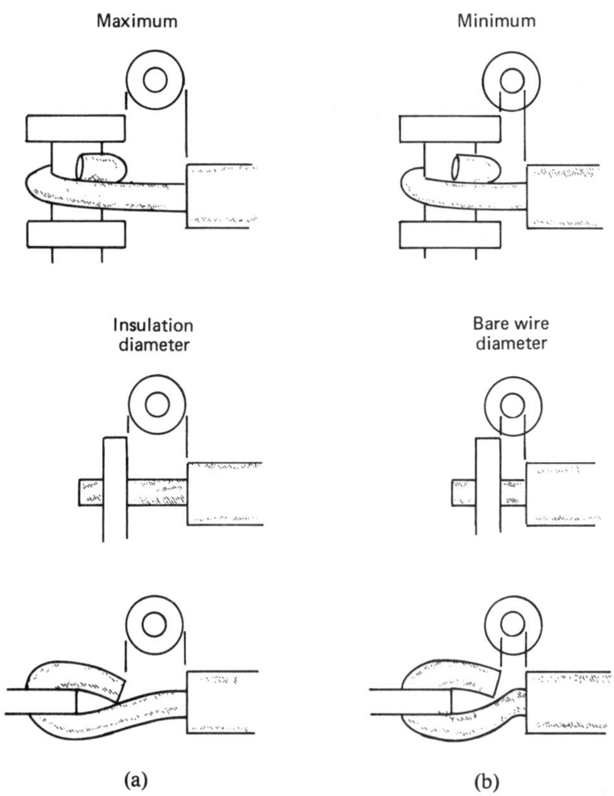

Figure 4–17 Proper strip gap. (a) Maximum gap. (b) Minimum gap.

ing a turret terminal, tinned wires that are AWG (American Wire Gauge) 26 and smaller should be wrapped more than one-half turn but less than one full turn around the post. Conductors that are larger than AWG 26 should be wrapped one-half to a maximum of three-fourths turn around the post (180° to 220°). All conductors must be confined to the guide post. More than one wire may be installed in a single slot (guide), provided each wire is wrapped on the terminal post and not on another wire. Also, the conductor must make contact with the terminal pad (Fig. 4–18).

When bending the wire to fit the terminal, use a rounded instrument or special round nose pliers (Fig. 4–19). If using conventional long nose pliers, use care to prevent nicking or damaging the wire. Leave the appropriate strip gap when placing the wire around the terminal.

Remember that for good bonding to take place there must be a proper application of solder and heat. Solder must be applied to the terminal on the side opposite the soldering iron (Fig. 4–20). The solder is melted by the hot terminal and flows around the wire to form a band. An excessive amount of solder is not used. The outline of each conductor in the wire must show through the solder and the surface must be shiny (Fig. 4–21).

4.8.2 Bifurcated Terminals

When soldering bifurcated terminals, the conductor can be routed either through the bottom or through the side (Figs. 4–22 and 4–23). When using the bottom route, the wire should not extend beyond the diameter of the base except as shown in Fig. 4–22c. When more than one conductor is connected to a terminal, the direction of bend of each additional conductor shall alternate (Fig. 4–23b). As with other terminals, apply the soldering iron to the terminal to the opposite side of the joint. A pool of solder may be added to the iron to help with heat transfer to the terminal.

4.8.3 Hook Terminal

Connections to hook terminals shall be as shown in Fig. 4–24. The bend to attach conductors to hook terminals should be a minimum of one-half turn to a maximum

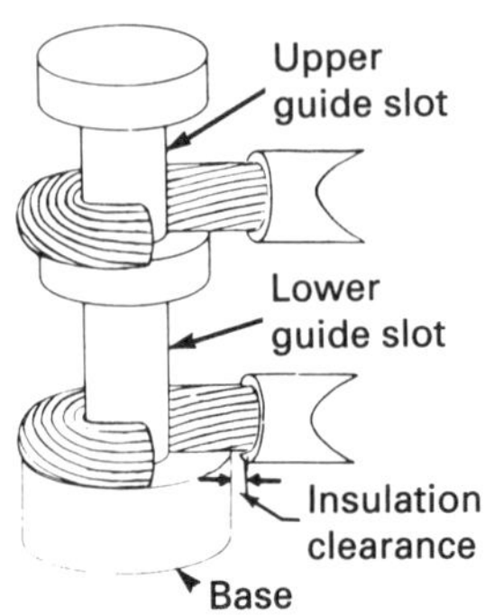

Figure 4–18 Turret terminal.

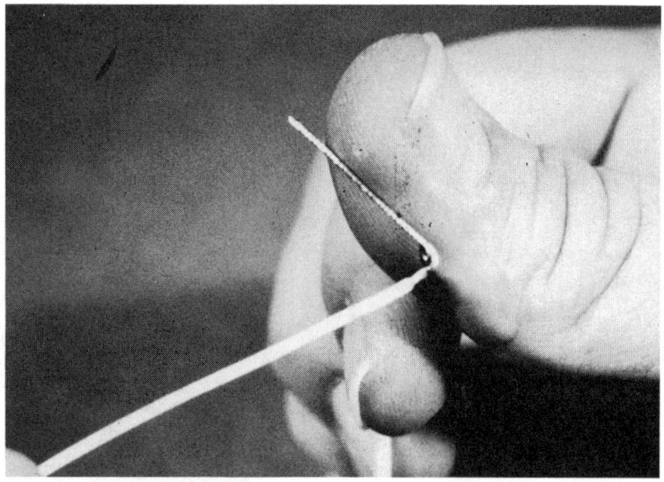

(a)

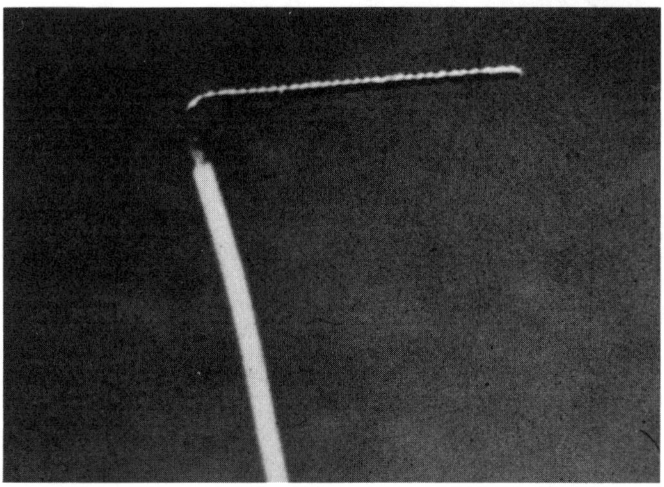

(b)

Figure 4–19 Preparing a wire to mount on a terminal. (a) Bending the wire. (b) Bent and tinned wire.

of three-fourths turn. Protrusions of conductor ends should be limited to avoid damage to insulation sleaving where it is used.

4.8.4 Solder Cup Terminal

The cup terminal is a hollow tube with a closed bottom. Conductors should enter the cup as shown in Fig. 4–25a. Conductors should be bottomed in the cup and shall be in contact with the inner wall of the cup. The maximum number of conductors should

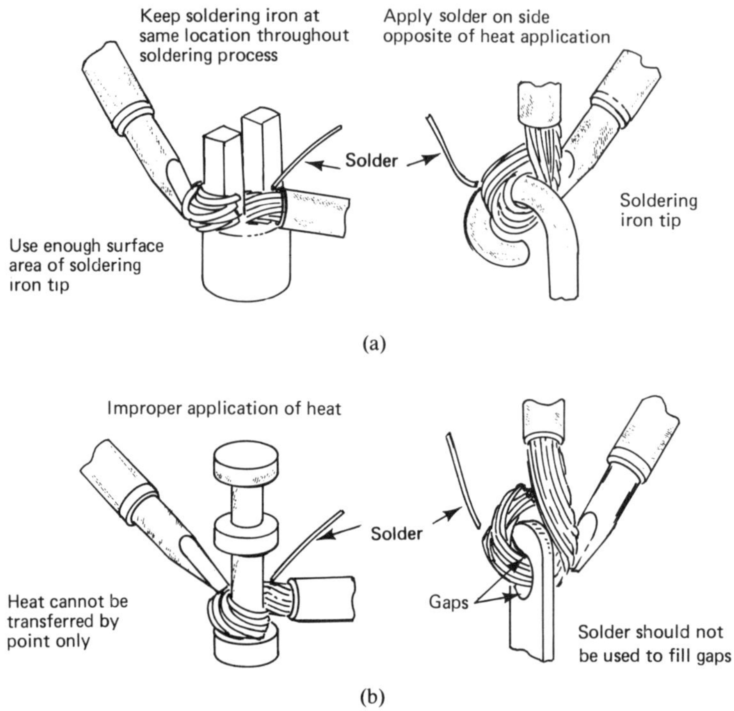

Figure 4–20 Proper placement of iron for a good bond. (a) Proper application. (b) Improper application.

be limited to those that can be in contact with the full height of the inner wall of the cup (Fig. 4–25a). To make a wire connection to a cup terminal, first strip and tin the wire. Fill the cup with solder. Reheat the solder, and insert the wire into the cup (Fig. 4–25b). Use an antiwicking tool to prevent solder from wicking under the insulation. Allow the cup to cool before moving the wire. The antiwicking tool also helps to prevent the wire from moving before the connection has cooled.

4.9 ACCEPTANCE AND REJECTION CRITERIA FOR SOLDERED TERMINALS

Acceptable solder connections are characterized by the following:

- Clean, smooth, bright, undisturbed surface.
- Contour of wire sufficiently visible to determine the presence of the wire, the direction of the bend, and the terminating end of the wire.

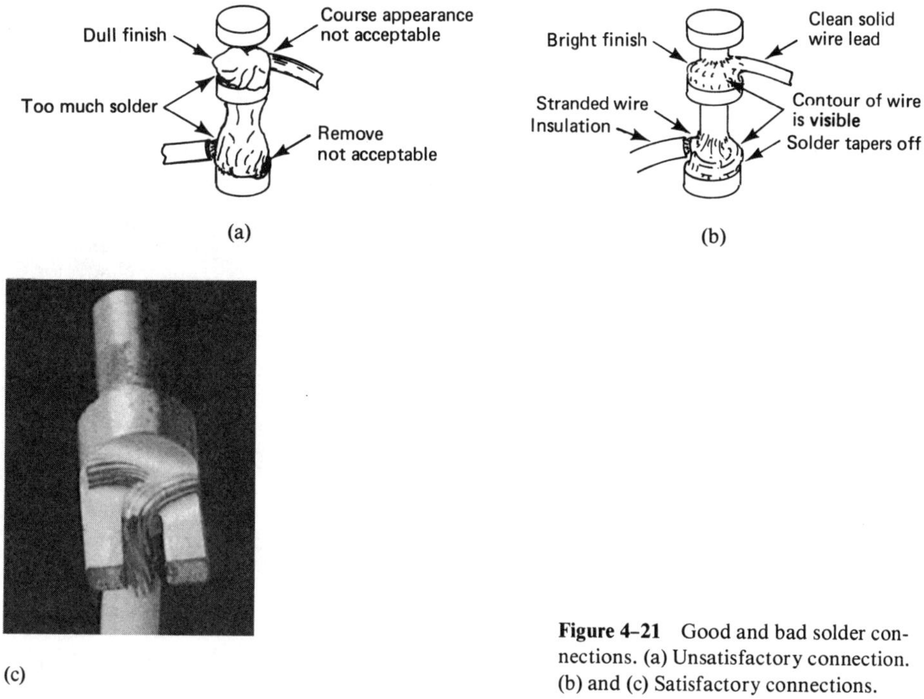

Figure 4–21 Good and bad solder connections. (a) Unsatisfactory connection. (b) and (c) Satisfactory connections.

- Complete wetting.
- Proper amount and distribution of solder and absence of defects.

The following are some characteristics of unsatisfactory solder connections, any of which is cause for rejection:

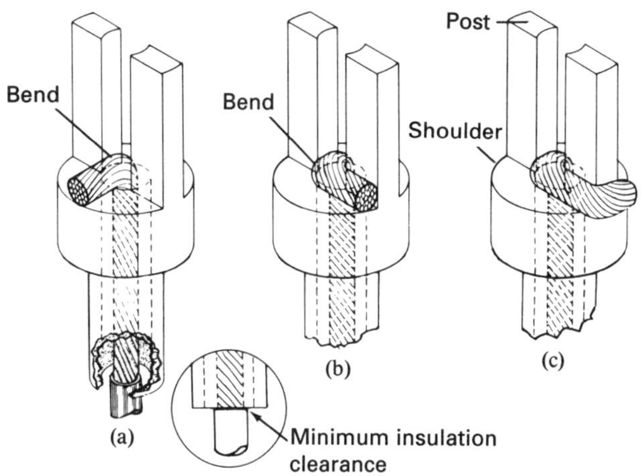

Figure 4–22 Bifurcated terminals: bottom route connections.

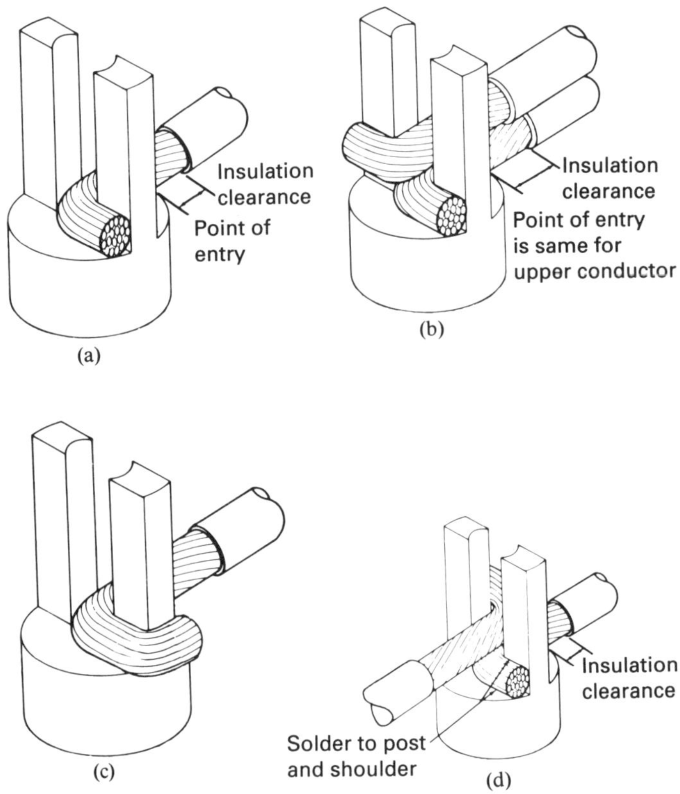

Figure 4–23 Bifurcated terminals: side route connections.

Conductors and Parts

- Damaged, crushed, cracked, charred, or melted insulation.
- Improper insulation clearance.
- Improper tinning.
- Separation of wire strands.
- Part improperly supported or positioned.
- Part damaged.
- Loose conductors.
- Cut, nicked, stretched, or scraped leads or wires.
- Flux residue or other contaminations.
- Improper wrap or stress relief.
- Spliced conductors.

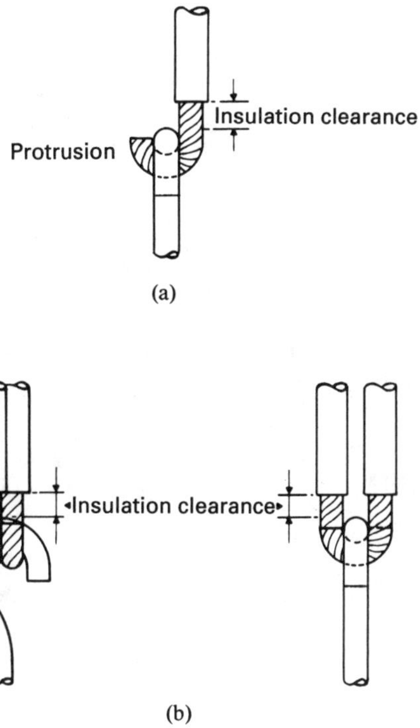

Figure 4–24 Hook terminal connections. (a) Single wire. (b) Multiple wire.

Solder Connections

- Cold Joint.
- Overheated granular joint.
- Fractured joint.
- Improperly bonded joint.
- Pitted or porous joint.
- Excessive solder.
- Insufficient solder.
- Splattering of flux or solder on adjacent areas.
- Rosin solder connection.
- Unclean connections (lint, flux, dirt, and the like).
- Dewetting, a surface in which the solder has not flowed evenly.

4.9.1 References Tables

Appendix G is a table of poor solder joints. It describes the defect, its appearance, the cause, and solution. Appendix H is a table of the nine elements needed to obtain a good solder joint.

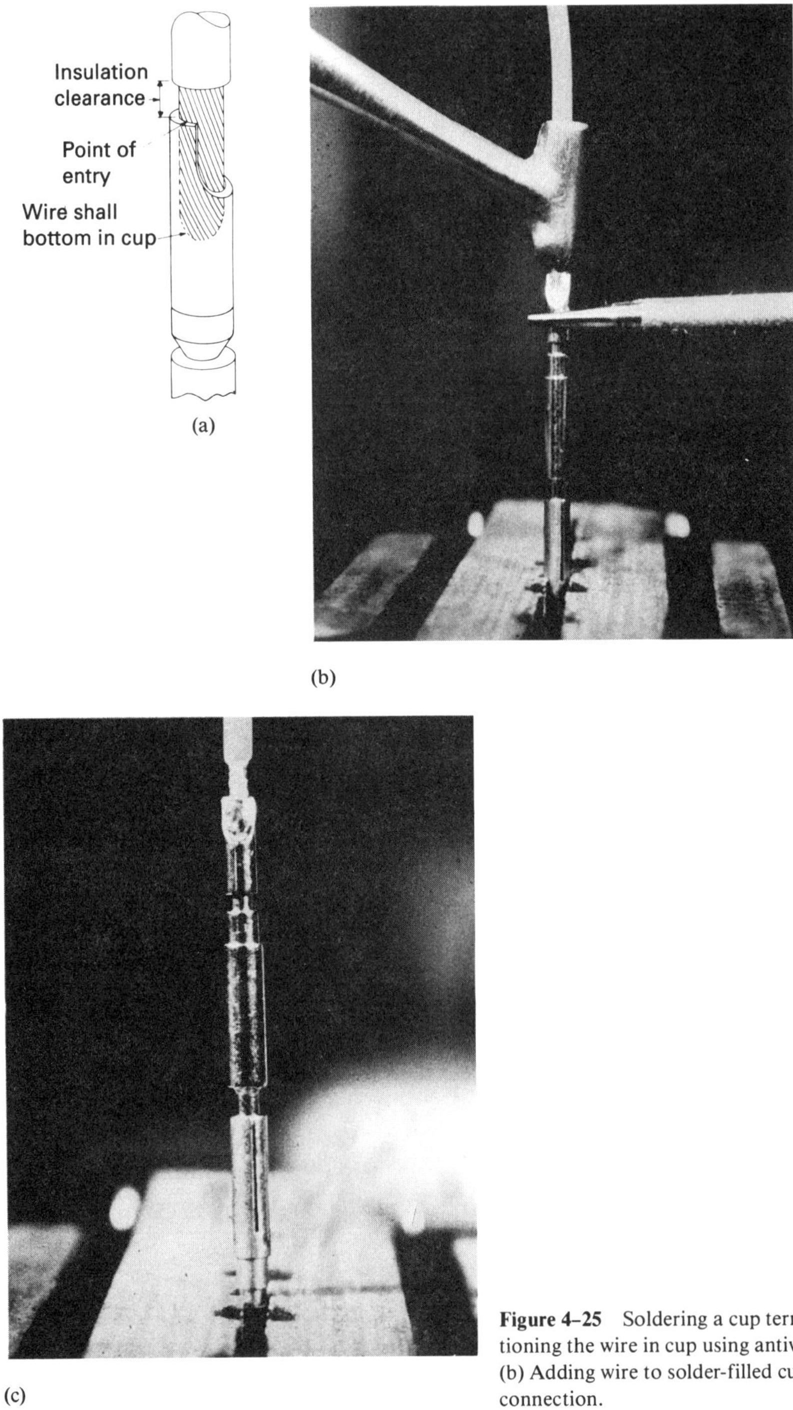

Figure 4–25 Soldering a cup terminal. (a) Positioning the wire in cup using antiwicking tool. (b) Adding wire to solder-filled cup. (c) Completed connection.

4.10 PRINTED CIRCUIT BOARD SOLDERING

There are two basic methods of soldering components to a PC board. One way is to mount all components on the board and solder them all at once by machine soldering (solder wave). This is typical of a production line where many boards are done at once. When only a few boards are to be soldered, the components are soldered on the assembly line. Therefore, the assembler can expect the soldering of a PC board to be one of his or her major soldering jobs.

Whatever method is used, the importance of high-standard workmanship cannot be overemphasized. Today, in the electronic industry, soldering is a critical skill. Engineers, electronic technicians, and electronic assemblers need to know how to solder correctly. This skill is necessary at all stages of the assembly and manufacturing process.

A typical work station includes a soldering iron (station), vise, liquid flux, isopropyl alcohol, 60/40 solder, acid brush, needle nose pliers, dikes, lead bender, component lead cleaner, cleaning cloth, wire strippers, other various components required for the job, and a PC board (Fig. 4–26).

Figure 4–27 shows both the component and trace side of a typical printed-circuit board. Components are not always mounted directly to a PC board. Specifications may require a component to be mounted to a terminal, *dip header,* or PC board eyelet (Fig. 4–28). The job specification sheet (work order) will call out the type

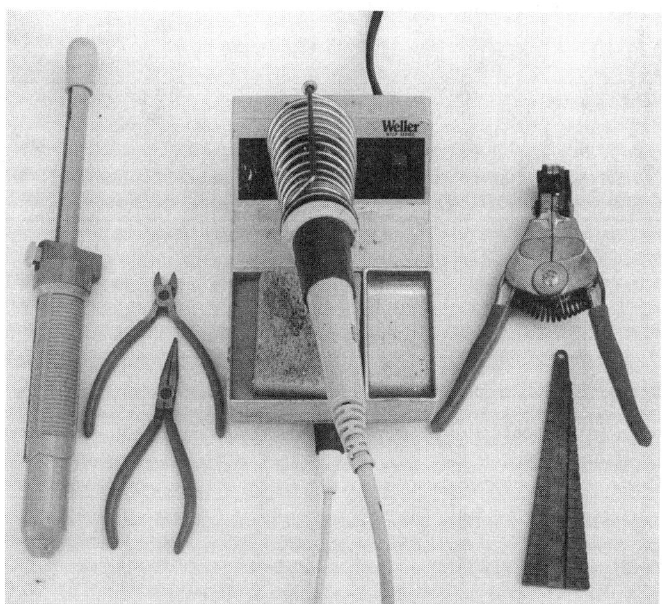

Figure 4–26 Typical work station. (Courtesy of Cooper Tools, Inc.)

(a) (b)

Figure 4–27 Typical PC board. (a) Component side. (b) Trace side.

of termination required. Whatever method of termination is specified, the same preparations are required for the soldering process.

4.11 SOLDERING PROCEDURE

- Prior to actual soldering, the components and wire to be soldered must be cleaned. If you clean the board chemically, handle it with gloves to prevent

Figure 4–28 Dip Header.

contamination from the oils on your hands. If it is not chemically cleaned, use a rubber eraser to clean the pad (Fig. 4–29).

- For NASA requirements, all component leads must be tinned prior to PC board mounting.
- Using a lead bender, prepare the component for insertion into the board (Fig. 4–30a).
- Mount the components on the printed circuit board (Fig. 4.30b).
- Place the soldering iron immediately adjacent to the point being soldered, and apply solder to the side opposite the iron (Fig. 4–31).
- Remember to provide good heat conduction from the soldering iron tip to the *trace* and component lead; a drop of solder may be added to the tip.
- The solder should just cover the lead so that the contour of the wire is sufficiently visible to determine the presence of the wire, the direction of the bend, and the termination end of the wire.
- The final junction should appear as shown in Fig. 4–32. There should be complete wetting of the junction; the pad is completely covered with the solder, forming a small cone up the edge of the lead.
- Remove excess rosin from the junction with an approved solvent (isopropyl alcohol or trichlorotrifluoroethane) (Fig. 4–33).

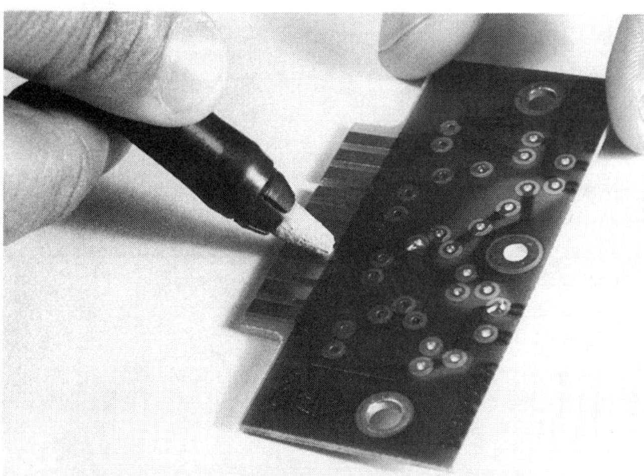

Figure 4–29 Cleaning a PC board with an eraser.

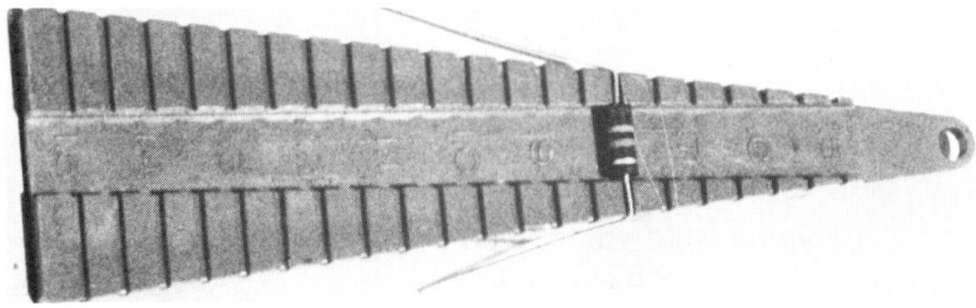

(a)

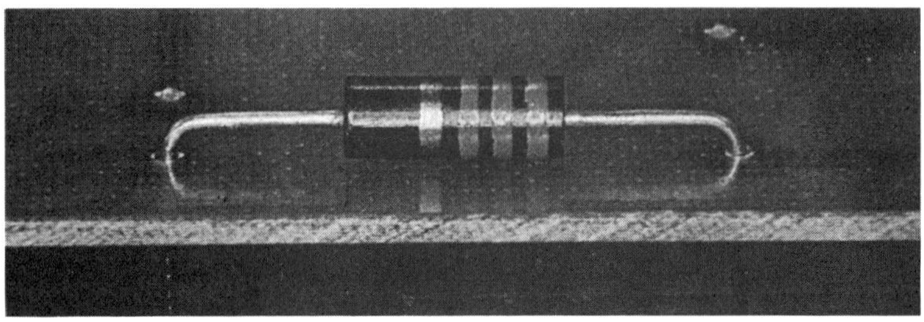

(b)

(c)

Figure 4-30 Preparing components for mounting. (a) Bending component lead. (b) Mounting component. (c) Mounting an IC using an IC inserter. (Courtesy of Micro Electronic Systems)

Figure 4–31 Placement of iron.

Check each soldered junction for unsatisfactory conditions (Fig. 4--34).

- Charred, burned, or melted insulation or parts.
- Burns on base materials (pads and the like).
- Bridging.
- Conductor pattern separation from board.
- Excessive solder.
- Flux residue, solder splatter, or other foreign matter on circuitry or adjacent areas.
- Dewetting.

Figure 4–32 Properly soldered joint and shape.

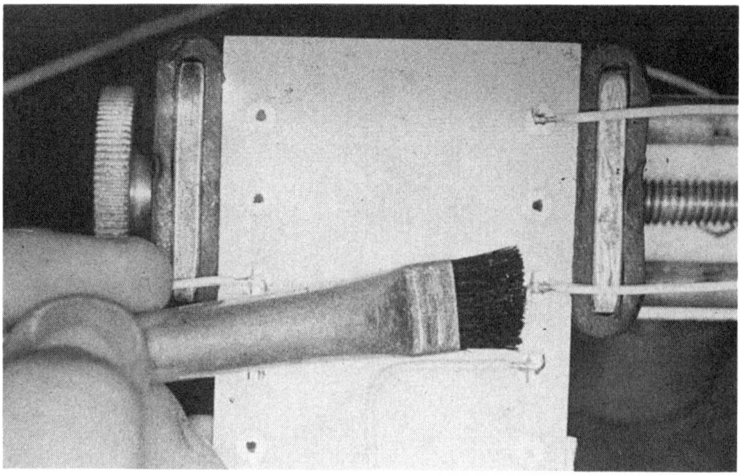

Figure 4-33 Cleaning flux from the joint.

- Insufficient solder.
- Pits, holes, or voids or exposed base metal in the soldered connections; icicles (cone-shaped peak or sharp points formed by removing the soldering iron too slowly); cold granular or disturbed evidence of grain structure change.
- Cut, nicked, gouged, or scraped conductors or printed wiring conductor pattern.
- Improper conductor length or direction of bend.
- Repaired or damaged printed wiring conductor pattern.

4.12 SOLDERING TERMINALS TO PC BOARDS

Use of terminals on a PC board should be restricted generally to situations where parts are expected to be removed and replaced several times or where there are other compelling design requirements.

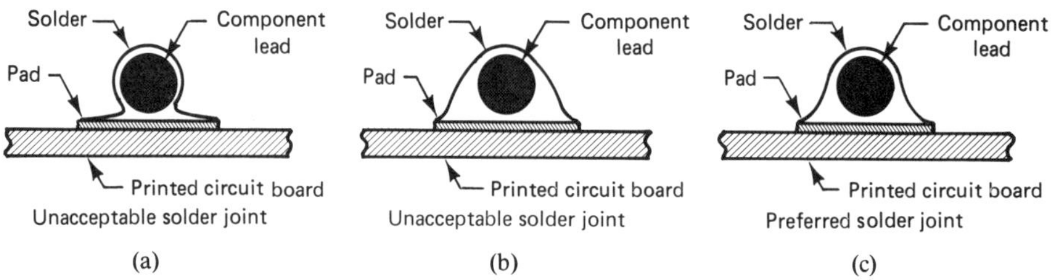

Figure 4-34 Cutaway view of solder joints. (a) Unacceptable: too little solder. (b) Unacceptable: joint has too much solder. (c) Acceptable: joint with proper amount of solder and heat.

Swage-type terminals have a terminal that is secured to a PC board by rolling or *swagging* the bottom of the terminal (Fig. 4–35a) and then soldered. A roll-type swage method is used on single-sided boards. An elliptical funnel swage (Fig. 4–35b) should be used for double-sided or plated-through boards. The process for mounting terminals to a PC board is the same as for soldering components to a PC board (Fig. 4–35c).

All swagged solder connections shall be heat sinked during the component lead soldering operation. Care should be taken to assure that swagged solder connections are not mechanically stressed during part-mounting operations (Fig. 4–36).

4.13 SOLDERING HEAT-SENSITIVE DEVICES

Many parts are sensitive to heat, particularly semiconductor devices. Heat sinks, which conduct heat away from the device being soldered, should be used where heat

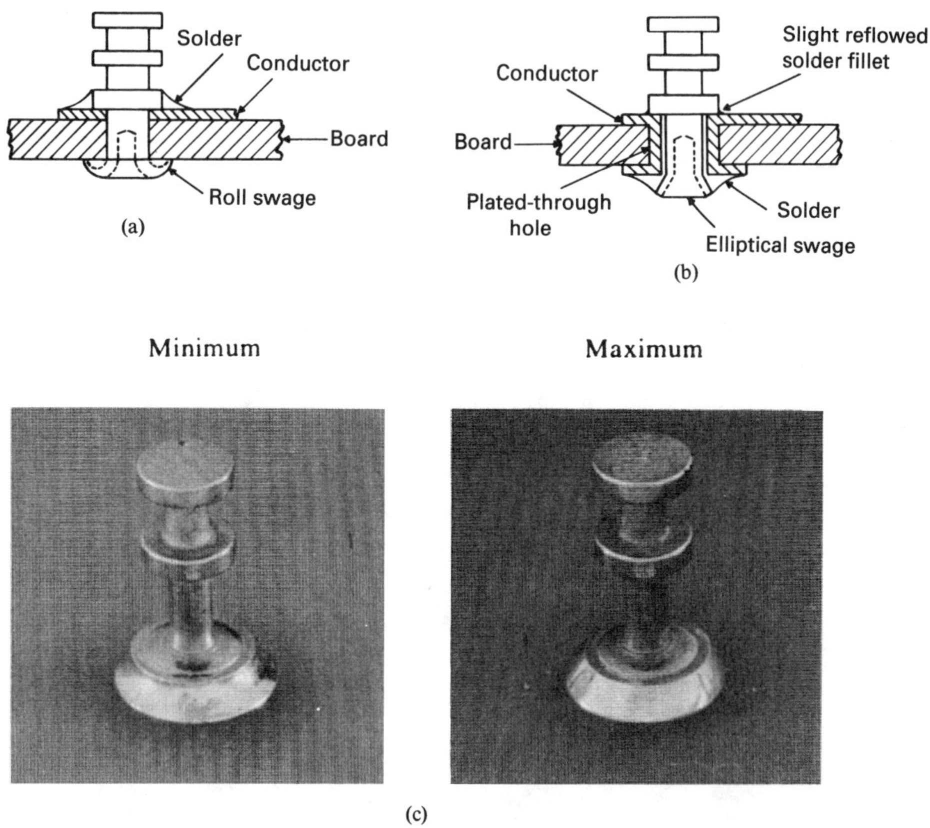

Figure 4–35 Mounting terminals to PC board. (a) Single sided. (b) Double sided. (c) Soldering of swagged turret terminal to pad.

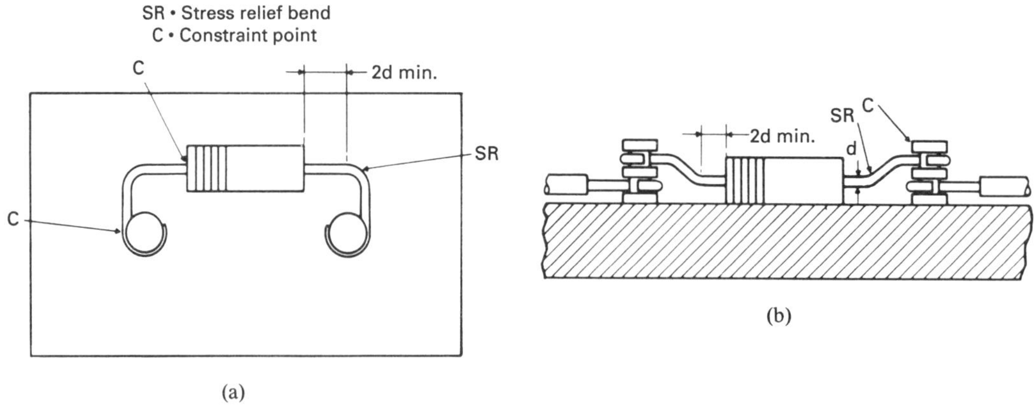

Figure 4–36 Components soldered to the terminal with stress relief. (a) Offset mounting. (b) In-line mounting.

during soldering operations may degrade conductors, insulation, parts, or a previously soldered connection.

Figure 4-37 shows heat sinks being used on each side of a heat-sensitive device. Heat sinks can be used in place of an antiwicking device when soldering conductors to prevent solder from wicking up under the insulation or damaging the insulation.

4.14 DESOLDERING OR REWORKING SOLDERED CONNECTIONS

Sometimes it is necessary to make changes to electronic equipment because of nonfunctioning parts, components installed incorrectly, or required circuit changes. This

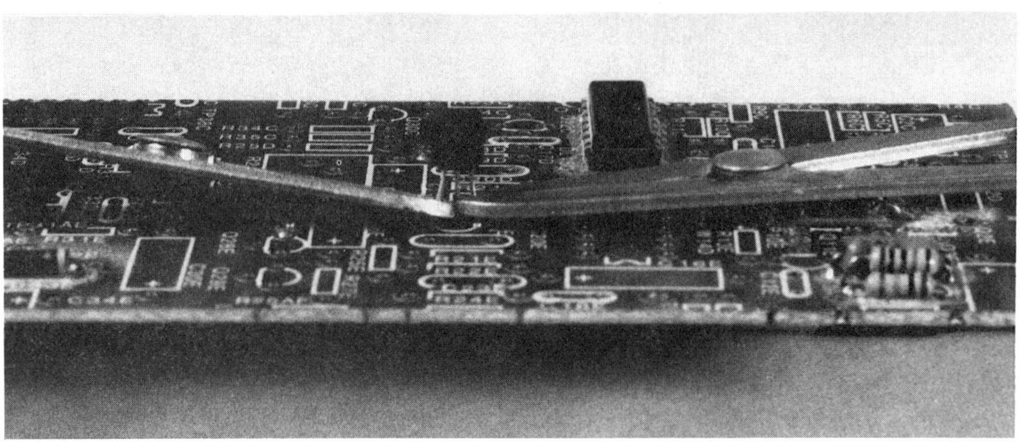

Figure 4–37 Protecting heat-sensitive devices.

process requires special care to prevent excessive heat damage to circuit components and PC assemblies. Care must also be taken to prevent excessive solder from flowing into components or across conductor paths.

The two most common methods for removing components from a PC board are using (1) a suction device or (2) a desoldering braid. Desoldering braid is a loosely woven stranded braid of copper wire. To remove solder from a component using desoldering braid:

- Coat the tip of the braid with liquid flux.
- Place the braid over the pigtail lead of the component and pad of the part to be removed.
- Place the tip of the heated soldering iron on top of the braid (Fig. 4–38).
- The heated braid melts the solder and the braid draws up the solder through *capillary action.*
- When the end of the braid is clogged with solder, cut off the clogged end.
- Clean the pad.
- Repeat these steps until the area is free of solder.
- Remove the defective or damaged part.
- Clean the lead of the new component.
- Form the leads to fit the holes, using a lead bender.
- Insert the new component and solder in place, using the techniques learned previously.

There are several types of suction devices for removing solder from a junction (Fig. 4–39). The process for removing solder is essentially the same, regardless of the type of device used.

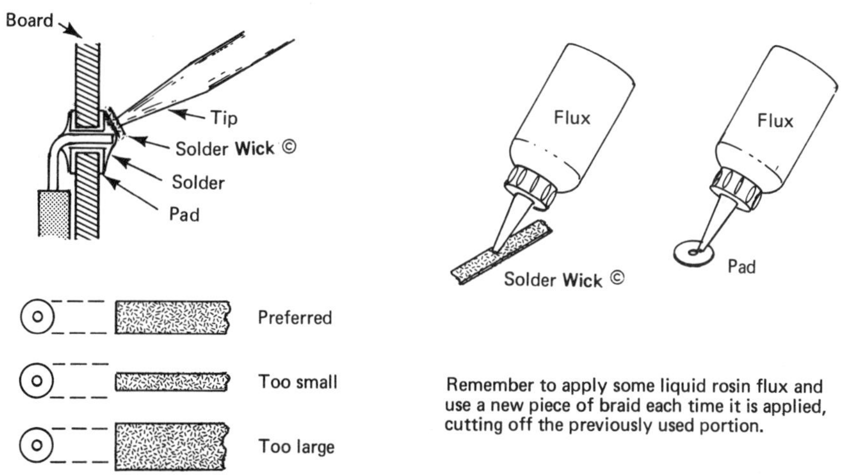

Figure 4–38 Using desoldering braid.

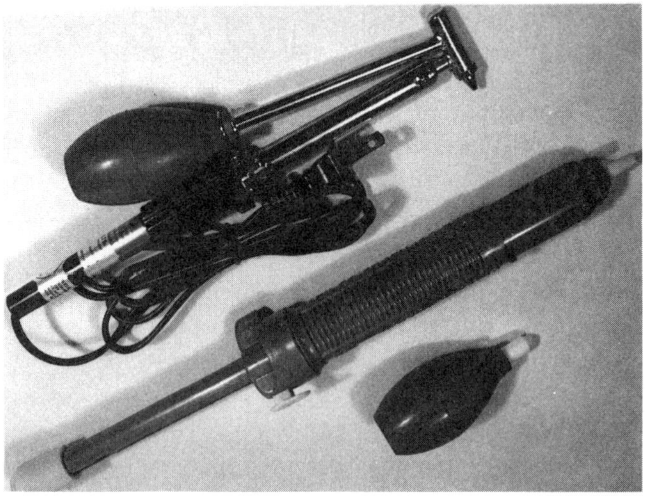

Figure 4–39 Desoldering tools. (a) Top-bulb type. (b) Center-solder sucker. (c) Bottom bulb on soldering iron.

- Heat the solder with the soldering iron.
- The heat-resistant Teflon tip of the solder sucker is applied directly to the heated solder, opposite the soldering iron.
- The solder sucker has a spring-loaded piston. Pushing the button on the side releases the piston (Fig. 4–40).
- The melted solder is sucked into the solder sucker by a vacuum action.
- This process is repeated until all solder has been removed (Fig. 4–40).

Sometimes it is difficult to remove a component. However, never use force or pressure to remove the part. If the ends have been bent or squeezed out of shape, cut them with a pair of dikes so that they may be removed from the board. If some solder remains on the component lead, apply heat and gently remove the leads before the

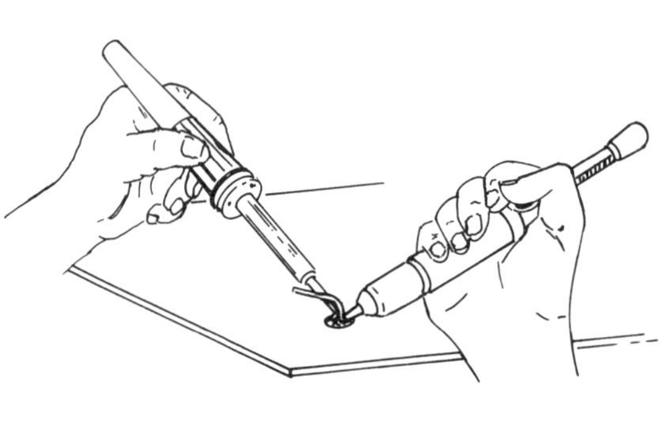

Figure 4–40 Using a solder sucker.

joint cools. It is important to pull the component straight up from the board without forcing or twisting it.

Integrated circuits (ICs) have multiple leads. The best method for removing an IC is to heat all terminals at the same time and gently remove the IC with a special "IC-remover" tool (Fig. 4–41b). Soldering irons have special tips for this purpose (Fig. 4–41a). After removing the IC, a solder sucker or desoldering braid can be used to remove the excess solder from the joint. An alternative method for removing an IC is to use a solder sucker or desoldering braid to remove excessive solder from each lead.

It is difficult to remove components from double-sided boards. The holes for the leads are plated through the board; therefore, solder adheres to the sides of the holes. To prevent damage to the board, it is recommended that a vacuum desoldering station be used. It contains a vacuum pump and has a special unit to control the temperature of the tip. Figure 4–42 shows the hand-held portion of the vacuum pump.

- Apply the tip to the board.
- Heat, using a back and forth motion. This causes the component lead to move when the solder melts.
- When the solder melts, depress the vacuum release control.
- The solder will be drawn up into the tip. This process is quick and exposes the board to a limited amount of heat.

Too much heat for too long may scorch the board or cause the pad or trace to lift off. When a trace is damaged on a PC board, it is not always necessary to discard the board. If a trace is lifted off the board, it may be repaired in one of two ways. *If the trace is not broken:*

- Clean the board and the trace.
- Apply an approved cement to the trace and board.

If the trace is broken:

- Cut off the portions of the trace that are charred or damaged.
- Clean the board with isopropyl alcohol. Make a new trace either with a piece of wire or with a copper strip the same width as the original trace (Fig. 4–43).
- Drill two holes of correct size near the trace, but not through it, to pass the wire through the PC board.
- Insert the wire and bend the ends over the existing trace.
- Solder the wire to the trace.
- This will give you a new conduction path, and the circuit will function as it did before the section of trace was damaged.

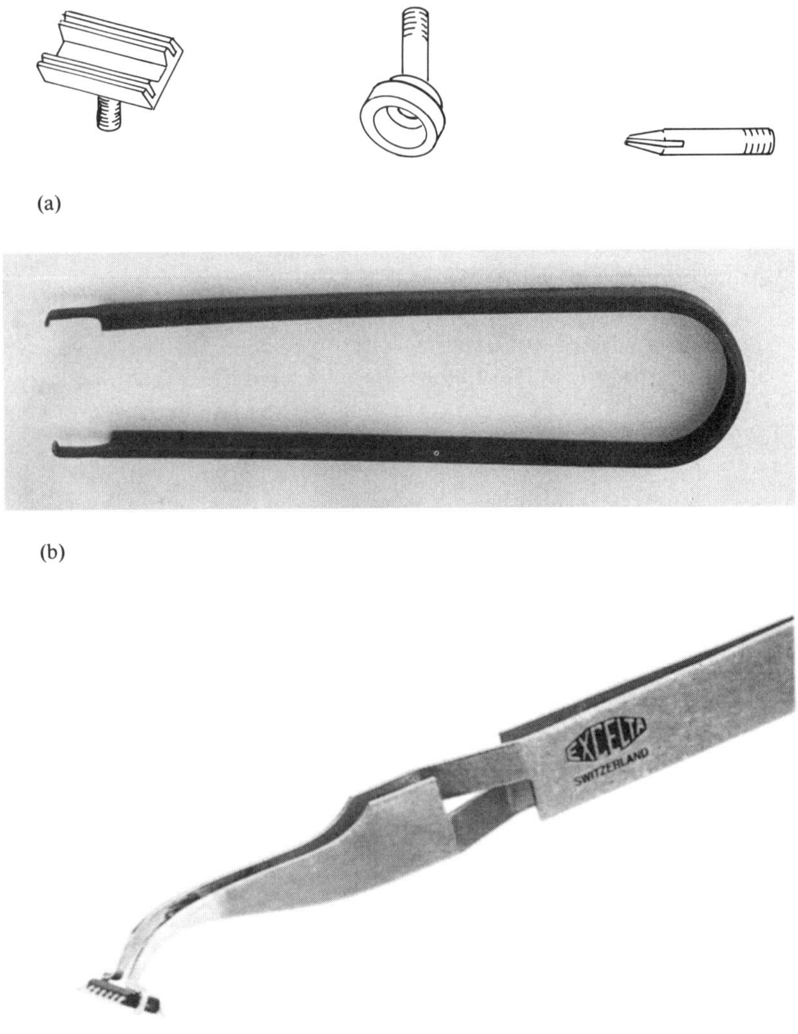

Figure 4–41 Special soldering iron tips. (a) Soldering iron tips for removing ICs. (b) IC remover. (c) Tweezers used for mounting surface-mounted ICs. (Courtesy of the Excelta Corp.)

4.15 INSPECTION AND REWORK STATIONS

Inspection of the PC board is an absolute necessity. Inspection should be performed prior to and after the assembly process. The continual change and complexity of PC technology creates problems in a manufacturer's ability to produce consistent preci-

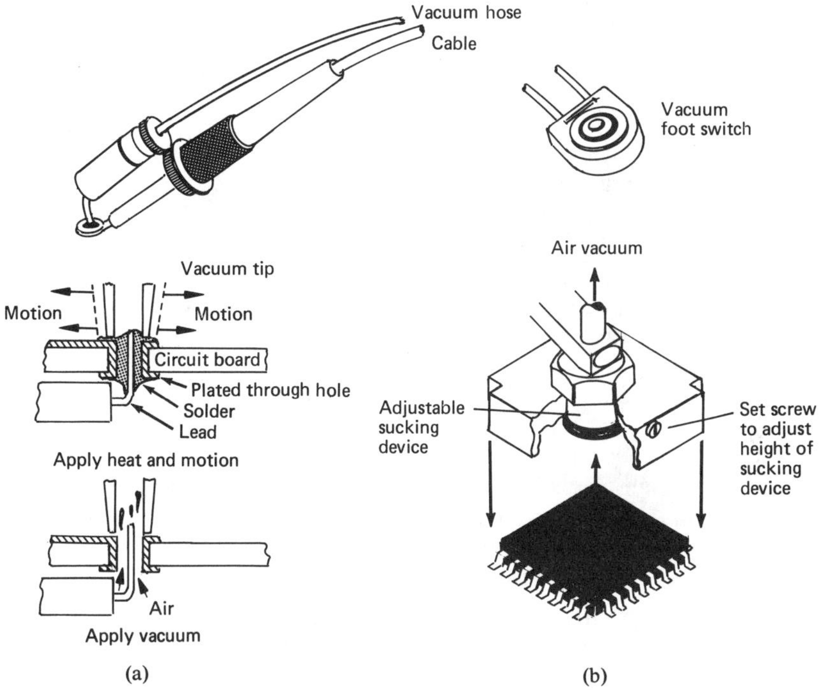

Figure 4–42 Desoldering double-sided boards. (a) Using a vacuum soldering station. (b) Vacuum soldering with SMD adaptor.

sion products with conventional automated soldering systems. Therefore, there is a demand for in-circuit testing, the testing of boards during the manufacturing process.

Traditionally, inspection is the last step of the manufacturing process. Corrections at this point are costly since the cost of reworking defective boards increases exponentially as the manufacturing process proceeds. For example, the cost index of in-process testing increases to 10 during system testing and to 100 when the defect is found after the assembly process is complete. Some new automated systems have been modified to perform inspection of soldering joints at the time they are made. Repairing a defective solder joint during the process takes about 30 seconds, compared to 3 hours for visual inspection and repair at the end of the assembly process.

The cost of testing is expensive, and adds only indirectly to the quality of the

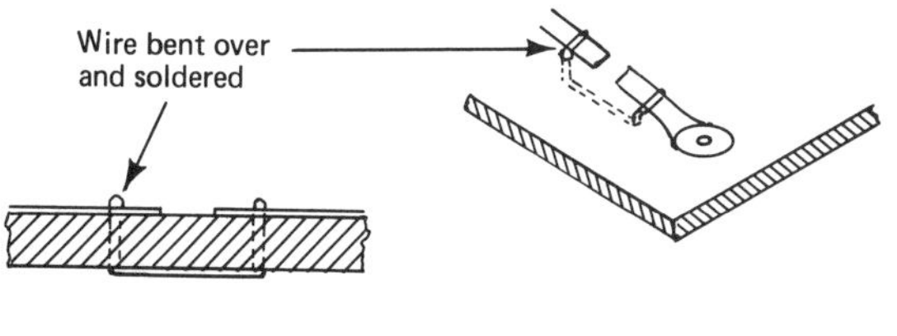

Figure 4–43 Replacing a broken trace.

product, but it is a necessary procedure for constant product reliability and quality. In the end, defect prevention is the ultimate in cost avoidance.

Even with in-circuit testing, PC boards require a final inspection. Inspections are performed visually with computers, and with x-rays at inspection stations. Originally, visual inspections were performed with a combined lamp and magnifying glass; now most visual inspections are done with a microscope.

Regardless of how automated the equipment is, some defects will occur, requiring rework and repair. Special rework stations are set up for this specific purpose. Depending on the skills and training of the inspector, he or she may perform this task.

The process of rework and repair requires special hands-on training. Training programs can be cost effective if done properly. Poor training is costly both in training time and product failure.

A rework station should include the following:

- A high-magnifying aid such as a stereomicroscope that will display a three-dimensional stereo image to aid the hand-eye coordination needed for rework. The most effective stereomicroscope is one that is inclined and adjustable to the operator (Fig. 4–44).
- A PC board positioning stage that will allow movement, including rotation of the board that is being inspected. Ideally, the stage would be foot controlled to free the operator's hands.
- The station should include ESD equipment to prevent damage to sensitive components. The ESD equipment should include such things as a conductive table top, conductive chair, foot and wrist straps, and proper grounding.

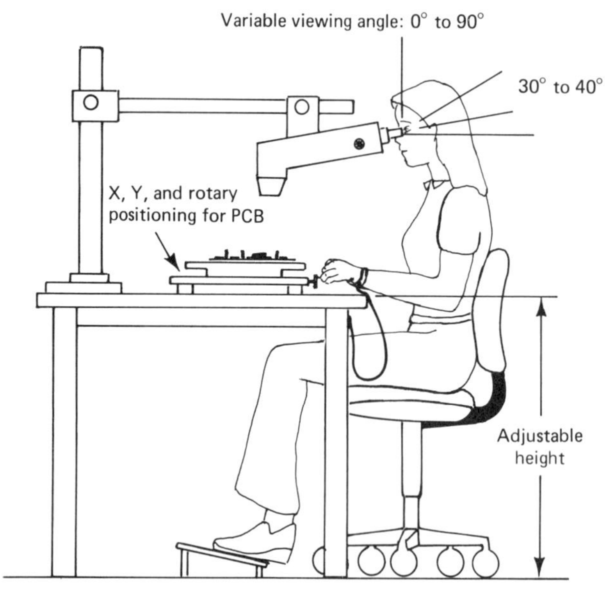

Figure 4–44 Example of a rework station using a stereomicroscope.

- Ideally, the station equipment would include computerized data-acquisition equipment, which would record and store information on defects such as defect location, defect type, defect disposition (how it was corrected), board ID, and operator performance.
- The acquisition of these data is important to the quality-control staff in the event of repetitive defects. It is important that the collection of the data does not interfere with the operator.

In case of a large operation, the process of defect location and correction might be videotaped for use in training classes.

4.16 SOLDERING SAFETY

1. Never hold solder in your mouth while soldering. Your soldering iron can give you a shock through this low-resistance path. Also, solder contains lead, which is poisonous.
2. Never test the heat of a soldering iron by feeling it with your hands; use a piece of solder instead.
3. Keep the soldering iron in its proper holder when not in use.
4. Place your soldering iron so that you do not have to reach across or around it while you are working.
5. When unsoldering a wire, make certain there is no tension or spring on the wire that could cause hot solder to be flipped into your face.
6. Never place rags, paper towels, or the like, near a hot soldering iron.
7. Wear safety glasses or goggles when soldering.
8. Use only a grounded-type soldering iron for both your safety and the safety of the circuitry on which you are working.

SOFT SOLDERING TEST

There are two choices for each question. Circle the correct answer.

1. Solder is a mixture of what two metals? (tin and copper *or* tin and lead).
2. What kind of joint results from moving the connection before the solder solidifies? (fractured *or* cold)
3. What solder is most often used in electrical work? (60/40 *or* 50/50)
4. In what state does soldering take place? (liquid *or* solid)
5. What is used to remove oil and grease from a joint before soldering? (solvent *or* flux)

6. Which could be used to remove oxides from a terminal? (alcohol *or* eraser)
7. What is used to remove flux residues from a soldered connection? (knife *or* solvent)
8. When soldering, which melts first? (solder *or* flux)
9. What must the assembler do before soldering to a PC board? (clean the leads *or* mount the component)
10. What must be done after the connection cools? (reflow the solder *or* clean the board)
11. What tool is used to select the proper distance between component leads? (soldering aid *or* lead bender)
12. How many times is the board cleaned? (two times *or* three times).
13. What can be done to help prevent lifting pads? (use extra solder *or* control the tip temperature)
14. What is one major job of an assembler? (making PC boards *or* soldering to PC boards)
15. Which method of soldering is generally used on the production line? (dip soldering *or* hand soldering)
16. In soldering, which is removed first? (heat *or* solder)
17. What are terminals used for? (end point *or* tie points)
18. What is done first when soldering wire to turret terminals? (clean the area *or* solder the joint)
19. What are terminals mounted on? (sheet metal *or* PC boards)
20. What does it mean when a wire is "tinned?" (solder is applied *or* tin is applied)
21. Which should be used when many connections are to be made? (turret terminal *or* terminal strip)
22. Which should be best to use if two connections are to be made? (turret terminal *or* terminal strip)
23. What is one danger to avoid when desoldering? (using too much heat *or* using too little heat)
24. Which tool is used to remove solder from a joint? (solder sucker *or* long nose pliers)

The following are true–false questions. Circle T for true and F for false.

T F **1.** When the heat is removed, the solder solidifies instantly.

T F **2.** The technique for soldering to a PC board is different than for soldering to a terminal.

T F **3.** After soldering, the joint is cleaned after it cools.

T F **4.** When soldering, the solder is applied before applying the heat.

T F **5.** Wires soldered to terminals need not be tinned.

T F 6. Sandpaper is used to remove oxides from terminal strips.

T F 7. All the old solder should be removed before soldering the new component in place.

T F 8. The best way to desolder an IC is to use a special desoldering tip.

T F 9. If a trace is broken, the board must be replaced.

T F 10. Loose pads may be cemented back in place.

Chapter 5

Automated Soldering Systems and Processes

5.1 INTRODUCTION

Automated soldering, as the name suggests, refers to the method by which components are mass soldered to a PC board. Prior to automated soldering, all components were mounted on the PC board and then hand dipped into a pot of molten solder (Fig. 5-1). The results of that method often produced uneven distribution of solder, unsatisfactory connections, bridges, icicles, and shorts. In many cases the connections could be repaired; however, the high volume of rework proved to be costly. Therefore, it became necessary to develop a more efficient and economical method for mass soldering. Automated soldering offered the opportunity for producing high-quality joints at a relatively low cost per joint.

There are many types of automated soldering systems on the market, but they all have certain features in common.

- Cleaner: cleans the board of contaminants prior to soldering.
- Fluxer: used for applying flux. This eliminates oxides in the solder wave at the PC board exit point to prevent solder bridges.
- Preheater: method of heating the board prior to soldering; it dries the flux and preheats components to reduce stress.

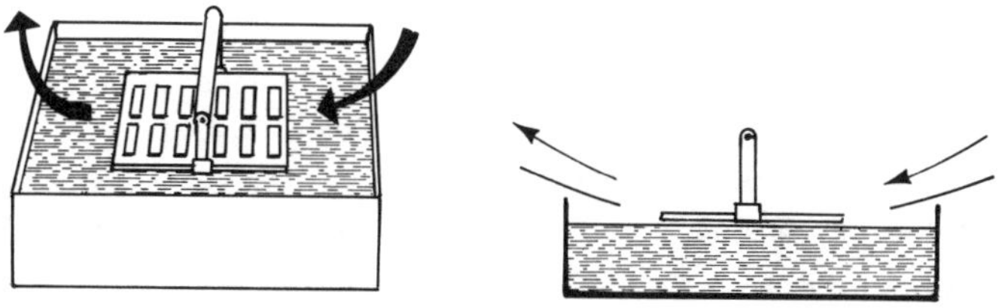

Figure 5-1 Hand-dip soldering.

- Solder pot: method of heating and holding the solder.
- Conveyer: method of moving PC boards through the system.
- Controller: method of controlling the temperature, amount of flux and solder to be applied, and speed at which the PC board will move through the automated system.

Originally, these systems existed as independent units with a conveyer belt between. Eventually, as the automated soldering machines became more sophisticated, they were integrated into one unit. This allowed for the addition of a better ventilation system, minimizing the problems of toxic fumes.

Later, extensive work was performed on the development of the actual wave (Section 5.2) in an attempt to improve the speed of soldering and reduce the problems of shorts, bridges, and icicles.

5.2 MACHINE SOLDERING

From the concept of hand dipping came the process of automated soldering. Solder was pumped through a slot to form a "wave" or mound of constantly moving molten solder through which the boards could pass. There were four major advantages to this method:

1. The surface of the solder was always clean and did not need skimming of contaminants.
2. The pumping action maintained a fairly constant temperature.
3. The raised wave provided a method for adding a conveyer system for automatically moving the PC boards over the solder bath.
4. The rapid movement of the PC board through the solder wave exposed the board to heat for a shorter period of time, decreasing problems of board warp and distortion.

Some systems that held brief popularity were the double wave, solder jet, and cascade. The cascade system, for instance, directed a stream of solder over a rippled bed to produce several waves in series on an incline conveyer (Fig. 5–2). Another system that held brief popularity was ultrasonics. The concept was to cause a *cavitation* at the joint, thereby avoiding the use of flux. But this system was costly and the soldering results were no better than with the conventional system. Also, ultrasonics are noted for causing damage to semiconductor devices. Therefore, today ultrasonic wave soldering is used mostly for tinning component leads prior to assembly.

Next came the introduction of oil injection. Oil on the surface of the solder wave reduces surface tension and oxidation and helps to drain excess solder from the PC board as it moves out of the wave. This aids in the wetting of parts, pads, and traces, thus reducing the incidence of icicles, shorts, and bridges. When using this system, it is necessary to consider the following criteria:

- Temperature stability and low evaporation loss.
- Length of "use life" before a change is necessary.
- Boiling point of material used.
- Wetting ability of the material used.
- Ease of removal from PC board assembly after the soldering operation.

As automated systems became more sophisticated, developers learned that the wave shape could affect the performance of the soldering machine and the quality of the soldered printed circuit boards. In the past, waves were parabolic or cone shaped and entered at a horizontal angle (Fig. 5–3a). This method is rarely used today. Boards left the solder region too rapidly, causing the joint to cool too quickly, resulting in the formation of short bridges and icicles. By using an angled conveyer and oil in the wave, surface tension was greatly reduced, allowing excess solder to drain more effectively from the soldered joint (Fig. 5–3b). Researchers discovered that they could process high-density board more effectively by using an asymmetrical wave. Since then, automated machines have been developed with variable wave forms and/or an-

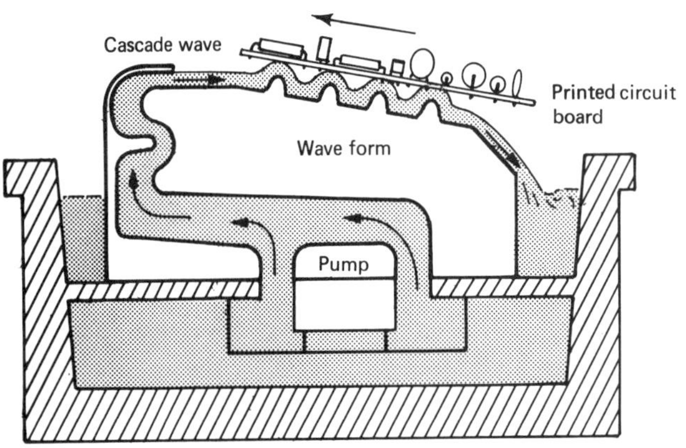

Figure 5–2 Cascade system.

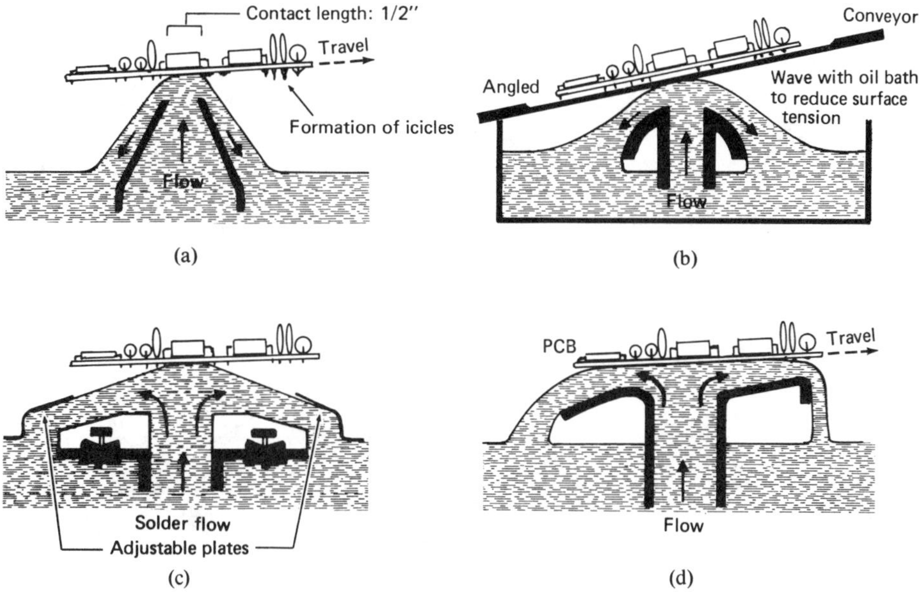

Figure 5-3 Variable wave forms. (a) Parabolic wave form: horizontal conveyor. (b) Parabolic wave form: angled conveyor for greater contact area. (c) Adjustable symmetrical wave. (d) PC board passing over an adjustable asymmetrical wave.

gled conveyors to meet specific needs (Fig. 5-3c and d). The static bath was one such design. This system, also called *drag soldering,* became popular, and many machines have been developed using this principle (Fig. 5-4).

Although *laser soldering* and *reflow soldering* are becoming popular, especially with *surface-mounted technology* (SMT), wave soldering continues to be the main soldering process in the assembly of PC boards. Although there has been a great increase of surface-mount technologies, the majority of PCBs use mixed technology, requiring both conventional leaded and surface-mount components.

5.3 PRESOLDER CLEANING AND PREPARATION

Prior to hand or machine soldering, PC boards and components must be properly cleaned to ensure a good-quality solder joint. PC board contamination results mostly during the process of fabrication: etching, cutting, drilling, and handling. These process contaminants (oil, grease, dirt, cuttings, and the like) can be removed with a vapor condensation of halogen hydrocarbon-type solvents. To accomplish this, the boards are placed in a vapor bath, thus forming moisture on the board surface. As the boards are slowly removed from the bath, the moisture vaporizes, removing the contaminants from the board. Because hydrocarbons cause air pollution, this type of process is generally not used today.

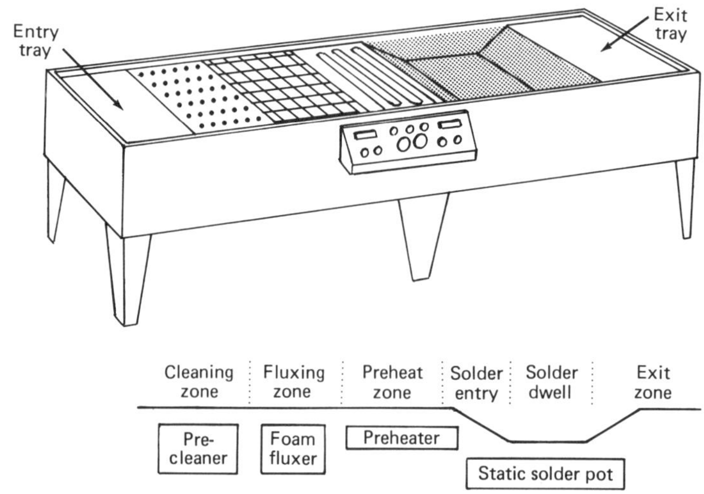

Figure 5–4 Drag soldering.

The preferred method for preparing PC boards and components for soldering is solder plating, that is, coating the surfaces with a thin layer of solder. The surfaces are then reheated (also known as *reflow*) to remove contaminants and uniformly distribute the plating over the surface of PC boards. This process eliminates the need to preclean PC boards and components with a chemical cleaner, and it is the most economical and reliable method for ensuring good solderability.

5.4 THE FLUXER

The automatic fluxer's main function is to consistently apply the correct amount of flux to a PC board prior to soldering to remove oxides and assure a good intermetallic bond. Several methods of applying flux have been developed to perform this function, such as brush, roller, foam, and spray. The brush method is one of the least efficient, because the brushes do not consistently apply an even coat of flux to the board.

The foam fluxer is a more efficient method. Compressed air is introduced into a porous air stone. As air escapes from the pores of the stone, it creates a foaming action in the flux. Foam fluxing is the simplest and usually the preferred method unless specifications prevent its use. Some considerations when choosing a fluxer are the following:

- Height of the flux (changes with board density and lead length).
- Matching the proper flux to the proper system.
- Monitoring flux density and specific gravity.
- Monitoring the temperature of the flux.

5.5 PREHEATERS

Soldering is only part of the process to obtain a good solder joint. The PC board is preheated prior to applying solder. This operation reduces the chance of thermal shock to components that would occur if the hot solder was applied to a cold board. This step also dries fluxes and presets them to their active stage, evaporates any remaining solvents, and allows an increased conveyor speed.

Types of preheating sources are cal rod heaters, quartz lamps, hot air blowers, and infrared heaters. These devices preheat PC boards to approximately 200°F.

Controlling PC board temperature can be difficult, since boards have different masses (number and spacing of components). A two-stage heating system, such as the one shown in Fig. 5.5, has helped to solve this problem. An infrared detector senses the temperature of the PC board at the end of the first heating stage. Then a microprocessor calculates the amount of added heat that is required of the second heater to bring the temperature up to the required 200°F. Whether the single- or two-stage heating system, temperature must be controlled to make the preheating state effective.

5.6 APPLICATION OF SOLDER

5.6.1 Solder Tank

The solder tank performs the wave soldering process of the automated soldering system. It contains the solder pot, the pump, and the nozzle that forms the solder wave. A turbulent jet of solder is pumped up into a nozzle, forming a head of solder through which the PC board is passed. The details of this process were discussed in Section 5.1. There are many styles and variations of the basic wave-form shape. The shape and angle of entry depends on the specific needs of the board, such as board size, number of components, and type of components being mounted.

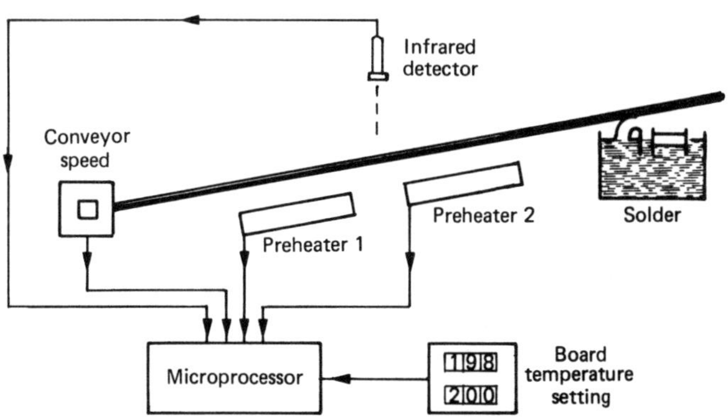

Figure 5–5 Two-stage automatic preheater.

Some considerations in designing the soldering stage of an automated soldering system are as follows:

- Size and number of joints to be soldered.
- Board composition, plating, and thickness.
- Component size and lead configuration.
- Will components be machine or hand mounted (assembled) on the PC board?
- Will the PC board have a single or mixed technology of conventional leaded and surface-mounted devices (SMDs)? SMDs are devices that are mounted on the surface of the PC board rather than using plated-through holes.
- Will SMD components be mounted on one or both sides of the PC board?

5.6.2 Solder Paste or Creams

When wave soldering is not possible, a process called reflow soldering is used. This requires an application of a solder cream or paste (a mixture of powdered solder and flux), which is applied to the joint to be soldered. Solder paste comes in various shapes and sizes, called preforms, and is automatically applied to the joint. Components are precoated with solder and then mounted to the PC board. Then the components are bonded to the PC board by the application of heat to *reflow* the solder. This process is used with heat-sensitive and surface-mounted devices. Reflow and SMDs will be discussed in detail in Sections 5.7 and 5.9.

5.6.3 Hand Soldering

Even the most sophisticated automated system will require a limited amount of hand soldering to:

- Remove bridges, shorts, and icicles.
- Remove defective or damaged parts.
- Replace components due to design change.
- Repair damaged traces.
- Add a limited number of SMDs.

5.6.4 Mixed Technology

The majority of PC boards processed today are and will for a long time be mixed technology, both conventional leaded and surface-mounted devices. This requires an added stage to the automated soldering system. The SMDs are mounted and soldered before the mounting and soldering of the conventional leaded devices (Fig. 5–6a). If SMDs are to be mounted on both sides of the PC board, an adhesive is used to hold the SMDs in place to prevent the parts from moving or dropping off during the sol-

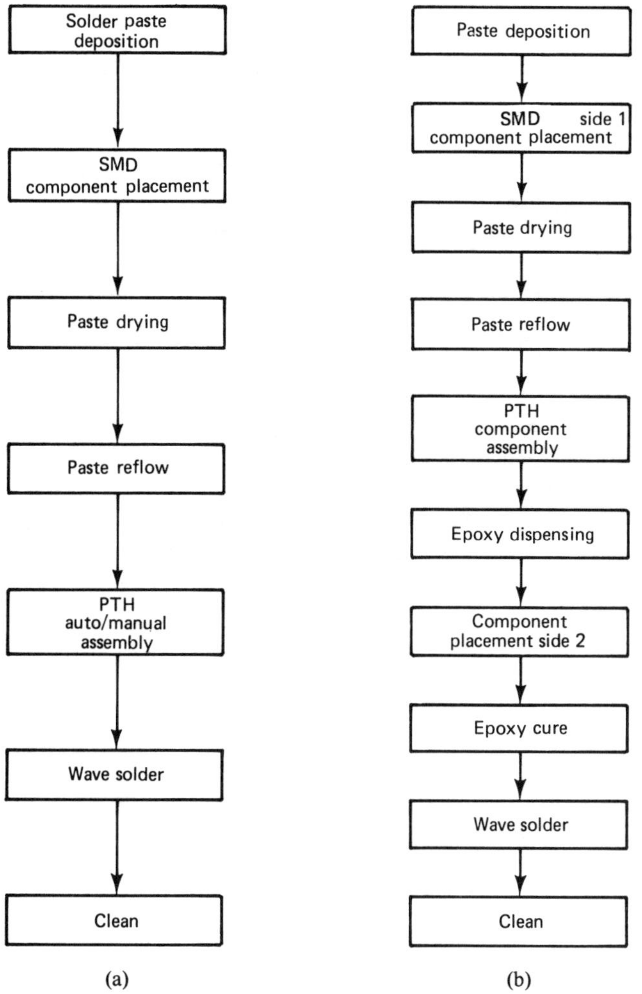

Figure 5–6 Mixed automated soldering. (a) Single-sided component mounting. (b) Two-sided component mounting.

dering process. Solder preforms are not used since the devices will pass through a solder wave (Fig. 5–6b). Problems with this, however, are unwanted solder (icicles) and the entrapment of flux beneath the devices, which could cause electrical failure.

5.7 REFLOW SOLDERING

Reflow soldering is the process of remelting solder on presoldered components and PC board surfaces (pad area) to form a soldered joint. This process requires several steps:

1. Presoldering: Coating the component leads with a layer of solder.
2. Applying solder preforms to the PC board surface: Generally, this is done automatically, but when a limited number of SMDs are added, this step is done by hand.
3. Paste drying: Solder paste preforms are dried to hold the components in place. This drying also prevents flux vapors from escaping too rapidly, causing fractured solder joints and moving and misaligning of components.

Heating methods such as vapor phase, infrared, resistance heating, and hot oil reflow can be used with this process. Presently, infrared and vapor reflow are the industry standards.

Some considerations for choosing a heating method are cost of heating PC boards to achieve soldering temperature, efficiency in heating the joint to soldering temperature, and durability of heat-sensitive devices.

Although laser and reflow soldering are experiencing increased popularity because of surface-mounted devices, wave soldering continues to be the main soldering process for PC boards.

5.7.1 Vapor Phase

A fluorinert liquid (fluorine base) is brought to the boiling point (419°F), forming a vapor blanket above the boiling liquid. As the PC board assembly enters the system, the vapor condenses on the surface of the boards, which quickly and evenly raises their temperature, providing a good solder joint (Fig. 5–7). This method is simple and clean, with a precise temperature-control system that protects most heat-sensitive components. The only problem is that all parts are exposed to the same temperature, which could cause failure to some semiconductor devices.

5.7.2 Infrared

An infrared beam can be focused on a very small area to heat only the joint to be soldered, without causing damage to components. This process can be expensive and

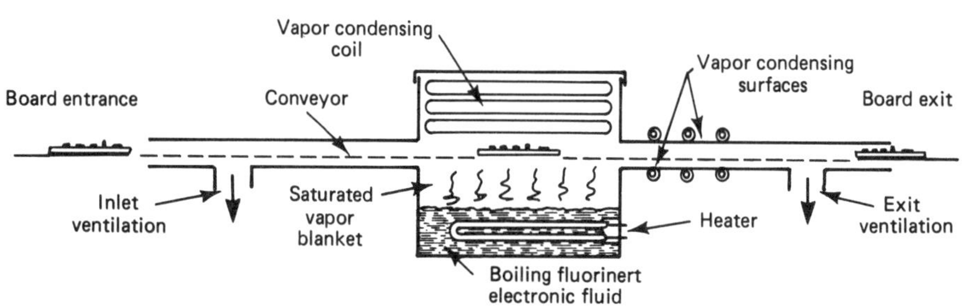

Figure 5–7 Vapor-phase soldering.

requires careful control to prevent damage to small PC board traces while attempting to heat the joint. On the other hand, large pads and conductor edges may not be covered with solder. The edges do not become as hot as the surface of the PC board due to the rapid speed past the infrared heaters.

Other types of infrared reflow use hot air to reflow the solder. With the *natural-convection* system, boards are passed through a heated chamber. Heat energy is transmitted to the PC board by convection and heat radiation from a natural draw of hot air from the entrance point through to the exit point. Another hot-air method uses forced-air convection. This method intentionally creates a movement of hot air-flow within the heat chamber with the use of blowers.

5.7.3 Resistance Heating

In resistance heating, current is passed directly through the parts or through a wire to heat parts. Heat is obtained from the resistance of the workpiece as current flows through it. This method is used for less critical processing since it is difficult to control temperature.

5.7.4 Laser Soldering

Laser soldering is one of the newest methods of applying heat to bond a solder joint. Lasers apply a very high energy beam, which is focused only on the solder joint being soldered. This method is fast; therefore, PC boards and components are spared from excessive heat, resulting in joints that are less brittle and more resistant to fatigue failure. Air cooling following the heating process in laser systems produces finer grain and stronger solder joints than other techniques.

Another added feature of laser soldering is that reflow soldering can be completed in one step. Components that could not withstand the massive heat used with vapor-phase or infrared soldering had to be added after this step. This created a problem because heating the added components also heated the existing solder joints, causing the existing surface-mounted components to move.

One disadvantage with laser systems is that they apply the same amount of heat regardless of the variation in joint size (dimension of pad and *land area*). Therefore, some laser joints do not reach the required temperature and this produces cold joints, while smaller joints receive excessive heat. This results in damage to components, leads, and PC board traces and pads.

A new system that has been developed to minimize this problem uses an infrared detector that senses (by heat) the size of the joint. The detector sends a signal to a computer controller to adjust the heat to the joint size.

Another laser soldering system that has been developed uses an infrared laser to inspect joints as they are being soldered. Repairs can be made in about 30 seconds, versus 3 hours for visual inspection.

5.8 THE WAVE OF THE FUTURE

Even with modern technology, there are still many problems with carefully controlled processing, such as wave height fluctuation, environmental changes, and the tendency of solder to go where it is not wanted. This can be seen in the amount of rework that must be performed after the soldering process. However, even with the existing problems of automated soldering, it is by far a better method than hand soldering or hand dipping.

The need to produce electrical units faster, more efficiently, and at less cost assures us that automated soldering is the way of the future.

5.9 SURFACE-MOUNTED TECHNOLOGY

Surface mounting is the method by which components are assembled onto the surface of the PC board. Components that are especially manufactured for SMT assembly are mounted onto solder pads called land areas (Fig. 5–8a, b). The pads and traces are extremely small, which makes component placement and alignment very difficult.

The traditional method uses plated-through hole (PTH) assemblies; component leads are inserted through the plated holes in the board itself. Since surface-mounted components are placed on the surface of the PC board, the handling of these boards becomes very critical to prevent the movement and misalignment of components. Ad-

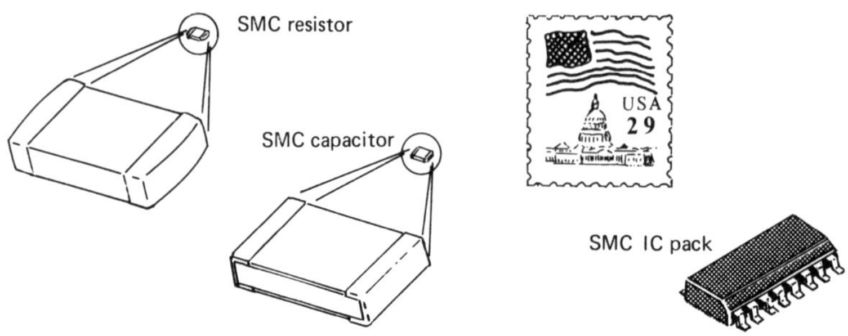

(a)

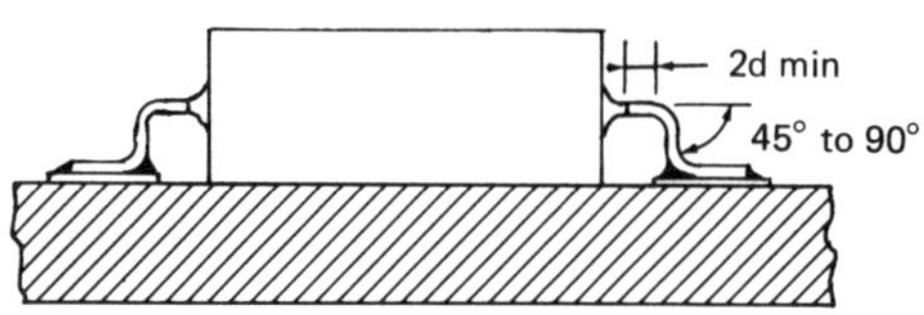

(b)

Figure 5–8 Surface mounting. (a) Surface-mounted components. (b) Surface-mounted board.

hesives are used to set and hold the components in place until the soldering process is complete.

The conversion from PTH to surface mounting is increasing rapidly, mostly because this method saves space, requires less materials handling, has improved shock and vibration resistance, has improved reliability, and is slightly more economical once initial manufacturing has been established.

With the advancement of SMT, the requirements for *in-circuit testing (ICT)* are becoming critical. ICT is a computer-controlled system that examines boards and components during production, rather than after (Fig. 5–9).

In the manufacturing of surface-mounted boards, the cost of reworking (repairing) defective products increases dramatically as the process proceeds. For example, if a defect is not found until after a product is in operation, it would cost 10 times

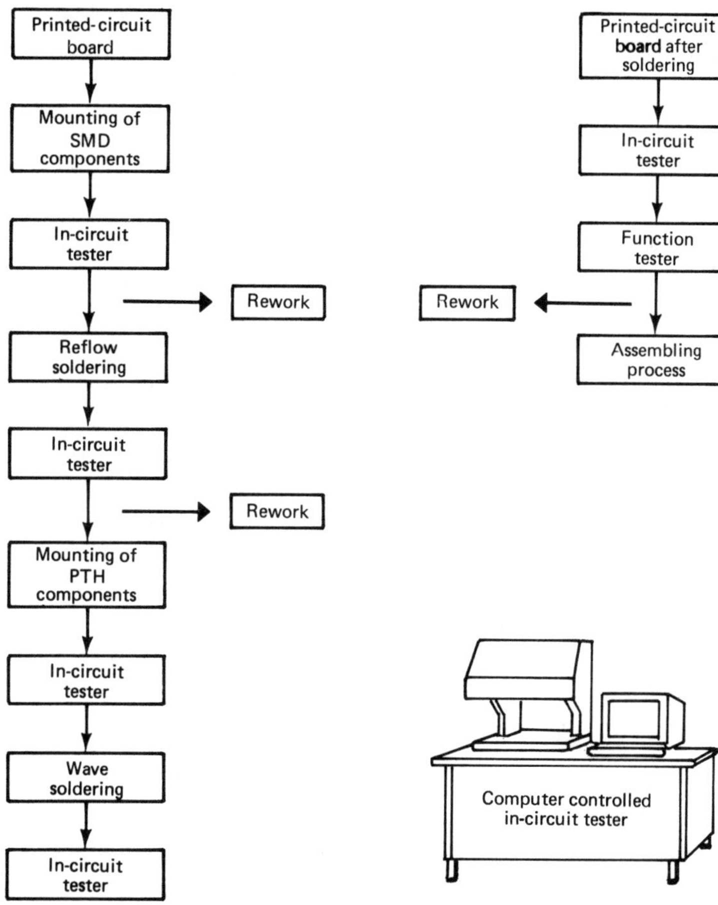

Figure 5–9 Computer-controlled SMT.

more to repair than it would during the manufacturing process. Therefore, it is crucial to try and eliminate defects at the earliest possible stage of the production process.

But it is not all good news. Presently, there are several disadvantages to using surface-mounting assembly. The specifications for surface technology are in a state of change. The trend has been toward smaller packaging of SMDs, and from plastic-leaded chip carriers (standard IC packages) to quad flat packs (QFD). In the past few years, industry has introduced even smaller QFDs known as fine pitch devices, that is, devices with even smaller leads and lead spacing.

This lack of standard in component size, lead length, and component thickness has created a problem for programming component pick and place equipments, PC board manufacturing, and automated testing systems.

An even bigger problem is the weakness of the surface-mounted joints themselves as compared with PTH joints. The shear number of joints on a board makes soldering the weakest link in the reliability of the finished product. Connectors create an even bigger problem, because of the stress to the joint from plugging and unplugging the connector. However, with the introduction of more sophisticated in-circuit testing equipment and better manufacturing procedures, this problem has been greatly reduced.

Other problems with surface-mount technology are as follows:

- SMT requires a large outlay for automated board handling.
- SMT requires more processes and therefore more machines than with plated-through hole technology, such as a screen printer, a glue dispenser, and a curing oven.
- SMT autoplacers are 20% to 50% more expensive than autoinserters for equivalent PTH machinery.
- PTH technology allows for greater flexibility in the packaging styles than does SMT.

The disagreement over standards is the major reason surface mount has not gained acceptance as rapidly as expected. There are currently at least three electronic industry standards: the Electronic Industries Association of Japan (EIAJ), the Joint Electronic Devices Council (JEDEC), and the National Semiconductor Corporation. The lack of standards for test procedures for equipment, components, and boards makes reliability data difficult to compare.

However, with the effort of a few major electronics companies, tremendous progress is being made both in mechanical and electrical standards for many types of surface-mounted components. Since SMT is still in its infancy, it is initially more costly than the PTH technology and has less industrial standards. As SMT becomes more widely accepted and standards improve, the technique will become fully automated. Increased production will result in reduced cost. It is expected that SMT will eventually replace PTH as the most desirable soldering technique.

5.10 ENVIRONMENTAL CONCERNS FOR MACHINE SOLDERING

Freon (CFC, chlorofluorocarbon) has been one of the most widely used solvents for cleaning (defluxing) PC boards. However, because of the growing concern over the effects of CFCs on the ozone layer and global warming (the greenhouse effect), industry is attempting to phase out the use of CFCs.

New solvents have been developed that have a much lower effect on the depletion of the ozone layer. Studies have shown that these new solvents have the potential ability to clean PC board assemblies as effectively as CFCs.

5.10.1 Process Improvements

The miniaturization of components and the popular uses of small-pitch IC packages (minimal lead spacing) tends to create a problem of solder bridging. Inert gas soldering, since it is a process that eliminates the use of fluxes, offers one solution. According to Sol Tech, Inc., fluxless soldering limits the risk of bridging because it avoids the possible cooling of solder at the exit side of the solder wave caused by outgassing (escaping of gas) of entrapped flux.

5.10.2 Elimination of Cleaning

To avoid oxidation of the solder wave at the board exit point, it is important to use flux containing activators. Flux must adhere to the board through the soldering process so that it is not washed off by the solder wave. Adhesion of flux and activators results in harmful residues after soldering; therefore, cleaning is required.

Board cleaning can be minimized by using newly developed *low-solid-content* fluxes that leave little residue. But they have drawbacks:

- Limited residues are still unacceptable to meet many MIL specifications.
- The activity of these fluxes is limited to a short dwell time (length of time heat is applied). The lack of adhesion caused by the washing effect of a jet wave generally leaves too little active flux at the exit point of the process, resulting in unacceptable solder joints.

The use of a spray fluxer and a single surface-mount solder wave application, such as Soltec's Smart Wave, can to some degree overcome these problems (Fig. 5–10). When the solder pot is placed in an oxygen-free environment, there is no need for flux. The solder wave will operate oxide free and, therefore, there is no need to carry the flux with the board over the full contact length, since the wave is clean.

Only a very small amount of preparation fluid (1% acid) has to be applied to clean the oxidized metal joint surfaces of the PC board. This fluid will be washed off by the solder wave, and acceptable postcleaning specifications can be met.

Figure 5-10 Automated soldering system. (Courtesy of Dover Soltec)

5.11 NEW TECHNOLOGY IN MACHINE SOLDERING

Currently, there are two types of machine concepts on the market: open and closed. The open concept, using flaps into a tunnel, will not reach the desired reduced oxygen rate by continuous flow of nitrogen alone. This system apparently needs the use of formic acid to actively reduce the oxygen level to create an inert environment.

This is a relatively simple machine, but the drawback of this concept is said to be the potential hazard of formic acid. Although used at low concentrations, it still needs special care. Safety standards in many companies prohibit the use of this material.

The Soltec closed-concept system is an inert gas machine that uses a tunnel that is absolutely sealed to the outside environment. The machine uses a vacuum chamber to accept boards from the outside. The vacuum chamber reduces the oxygen level. The boards enter into a completely sealed tunnel, kept *inert* by a small nitrogen overpressure. After preconditioning and wave soldering, the board leaves the machine again through a vacuum chamber, allowing a controlled inert atmosphere without the use of hazardous acids. Some advantages to this system are as follows:

- Cycle time appears to be around 30 seconds or less.
- The acceptance of component height up to 100 millimeters without excessive extra use of nitrogen.
- Significant cost reductions because there is no solder waste through oxidation (dross) and limited maintenance (no flux or dross, no cleaning of pallets, and so on).

MACHINE SOLDERING TEST

The following are true–false questions. Circle T for true and F for false.

T F **1.** All automated systems have certain features, such as cleaner, fluxer, preheater solder pot, and controller.

T F **2.** The original automated soldering system existed at one large unit.

T F **3.** A solder wave is a constantly moving mound of solder.

T F **4.** Ultrasonics is popular because it can be used with semiconductors without causing damage to the devices.

T F **5.** One disadvantage to automated soldering is overexposure to heat compared to hand soldering.

T F **6.** Wave shape does not affect the quality of the soldered PC board.

T F **7.** The brush method of applying flux is the most efficient method used.

T F **8.** Vapor phase is one method used for heating solder prior to soldering PC boards.

T F **9.** Resistance heating is the least preferred method of preheating because it is difficult to control temperature.

T F **10.** Laser soldering is used mostly for plated-through whole processing rather than surface mount.

T F **11.** Surface-mounted assemblies are mounted onto solder pads called land areas.

T F **12.** Inert gases are used in an environment that increases oxygen content.

T F **13.** Low-solid-content fluxes leave less residues.

T F **14.** An advantage of an open concept for automated soldering is the use of formic acid.

T F **15.** There are no limitations on the use of Freon (CFC) for the cleaning of fluxes from PC boards.

Chapter 6

Wire Wrap

6.1 INTRODUCTION

Wire wrapping, also called solderless wrapping, is the technology of connecting solid wires tightly around terminals, without soldering. With this method, wire is wrapped around a metal post (stud or pin), which is mechanically and electrically secure. The pressure of the wrap cuts through the oxidation layer on both the wire and terminal (Fig. 6–1).

In the past bread boarding was the preferred method of terminating wires for prototype, test fixtures, and nonproduction units. With today's smaller complex circuits and high-density boards, there is a need for a faster and more reliable method of termination. Wire-wrap systems can make approximately 300 to 800 wraps per hour and can be computer controlled, eliminating pin selection errors.

Originally, wire-wrap connections were done manually or with automatic hand-held tools (Fig. 6-2a, b, and c). Although this method is still performed on smaller boards and some prototypes, it is too slow for today's technology. Now, most wire-wrap processing is performed by automatic or semiautomatic machines (Fig. 6.2d and e). The new automatic autowrap machines work with CAD systems, making the wire wrap process faster and more accurate.

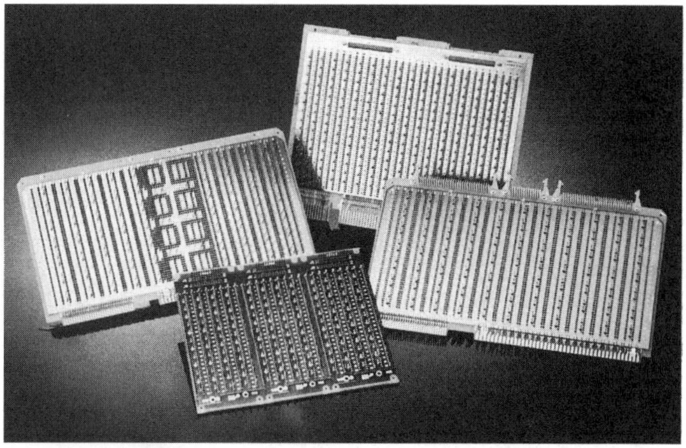

(a)

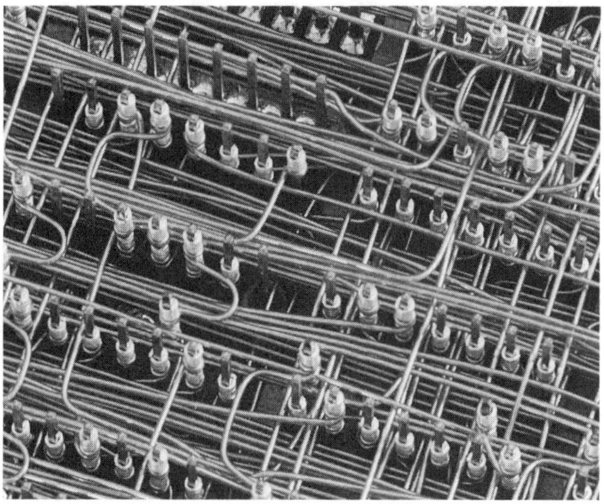

(b)

Figure 6–1 Wire-wrap board. (a) Before wiring. (Courtesy of dataCon Interconnect Sub-systems) (b) Completed board. (Courtesy of dataCon Interconnect Sub-systems)

6.2 MANUAL WIRE WRAP

Manual wire wrap is only used in cases were there are relatively few wraps or connections. Although the tools are inexpensive, it is too slow for most work.

The manual tool uses a larger-gauge wire with a different type of insulation than the automatic tool (Fig. 6–3). Hand wire wrapping can only be used with terminals that will accept a connection from a hand-held tool.

For manual wire wrap, the end of the wire is not stripped prior to wrapping. A special tip strips (cuts) the insulation as it is wrapped around the post. The wire must

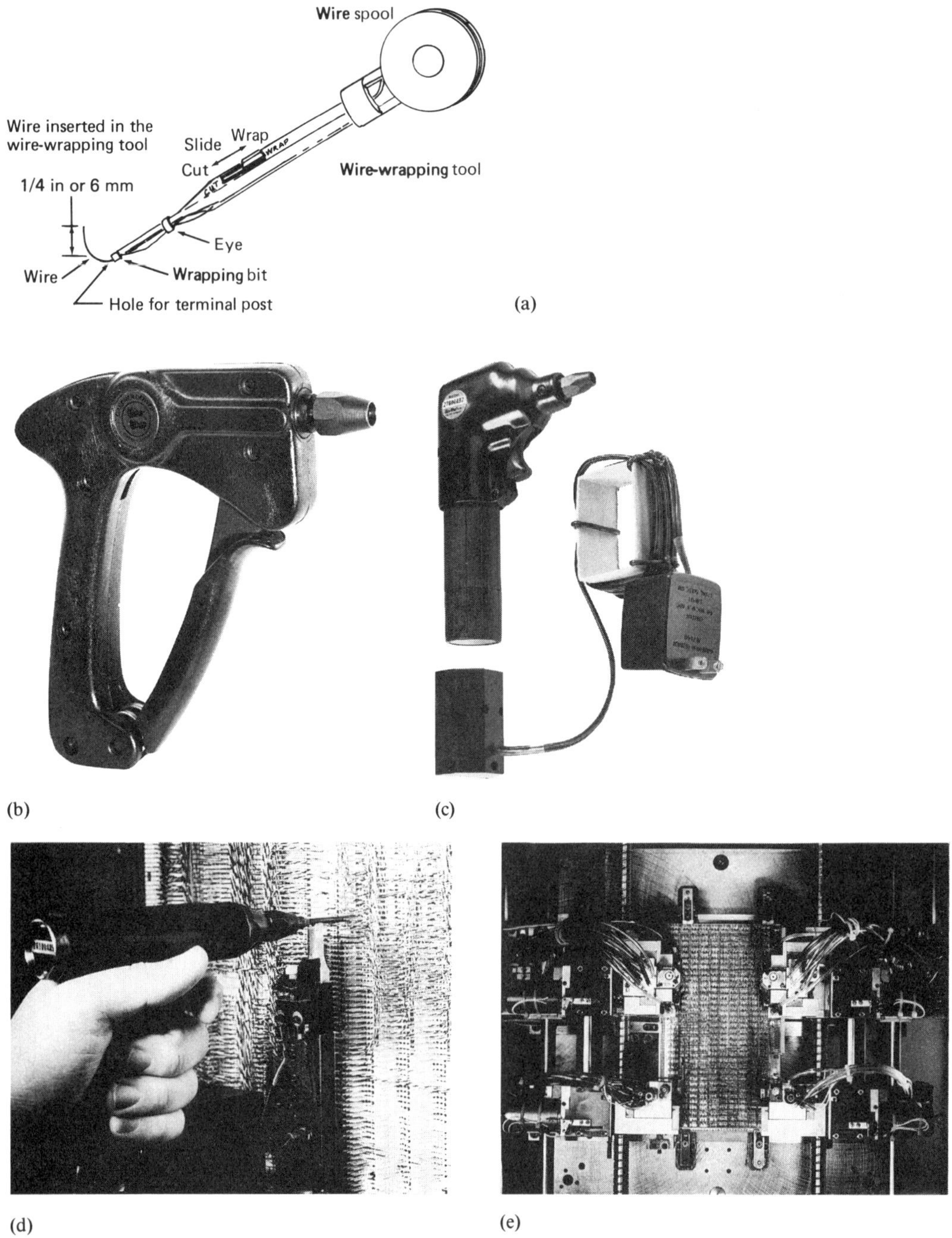

Figure 6–2 Wire-wrap tools. (a) Manual wire-wrap tool. (b) Electric hand-held tool. (Courtesy of Cooper Tools, Inc.) (c) Battery-operated hand-held tool. (Courtesy of Cooper Tools, Inc.) (d) Semiautomatic wire-wrap systems. (Courtesy of dataCon Interconnect Sub-systems) (e) Automatic wire-wrap system. (Courtesy of dataCon Interconnect Sub-systems)

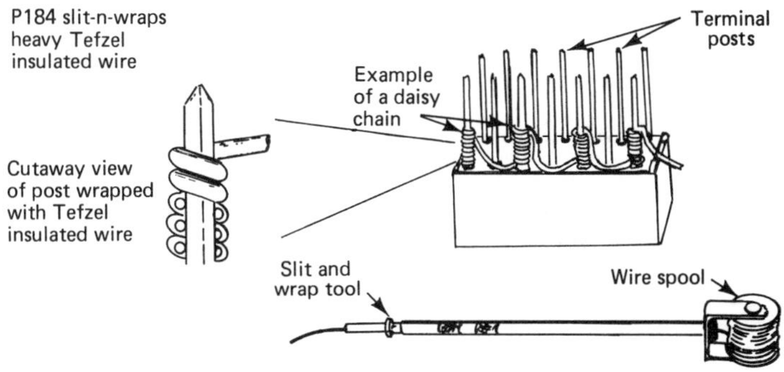

Figure 6–3 Manual wire-wrap tool and wire-wrap connection.

be wrapped in a clockwise direction or the cutter will not function. A counterclockwise wrap results in an unacceptable connection. The procedure is as follows:

- Set the slide in the wrap position
- Use a firm, but gentle pressure while turning the tool. The wire is automatically pulled down from the spool while making the wrap.
- Lift the tool from the first terminal and move it to the next. The tool makes one continuous wrap from each point to point connection (also called a daisy chain).
- When the connection is complete, move the slide to the cut position.
- Turn the tool approximately one-half turn in a clockwise direction. This will cut the wire close to the terminal.

6.3 AUTOMATIC HAND-HELD WIRE-WRAP TOOLS

There are three types of automatic hand-held tools: pneumatic (air), electric, and battery operated. A pneumatic tool is preferred for production work, but an electric tool can be used if compressed air is unavailable. Battery-operated tools are generally used for smaller jobs or for repairs and modifications.

Unlike the manual wire-wrap tool, in which a continuous wire is used from point to point, the automatic hand-held tool requires that each wire be cut to length for each point to point connection. The wire-wrapping process will be explained later in the chapter.

6.4 MATERIALS AND TOOLS

6.4.1 Wire

Wire-wrap wire is a solid wire made of a strong copper alloy to minimize breaking or cracking when unwrapping. It is coated with tin, a tin–lead alloy, gold, or silver.

Standard wire gauge sizes are nos. 18 to 30 AWG; nos. 26 to 30 AWG are used most often.

6.4.2 Terminal or Post

Terminals or posts are usually square, rectangular, flattened, or V-shaped. They range in length from 0.3 to 0.6 inch. The side dimensions are generally 0.025 to 0.045 inch. Each post can accept two or three adjacent connections, depending on its length.

6.5 CLASSES OF WRAP

There are two classes of wire wrap: standard (or class 1) and modified (or class 2). If not specified in engineering drawings, the modified wrap is the preferred method, with standard wrap generally accepted for nos. 18 through 24 AWG wire.

A standard wrap produces a wrap with only the bare wire around the terminal post (Fig. 6.4a). This type of wrap is used in equipment that is stationary.

A modified wrap produces a wrap with a portion of the insulation around the terminal post (Fig. 6.4b). This increases the connection's ability to withstand vibrations. This type of wrap is used in equipment that is subject to vibrations from frequent transportation, such as those found in airplanes, ships, and rail systems.

6.6 HOW TO MAKE GOOD WIRE-WRAP CONNECTIONS

6.6.1 Wrapping Process

During the wrapping process, the wire is pulled from the wire slot by the rotation of the bit around the terminal. While the wire is wrapped (under tension) around the

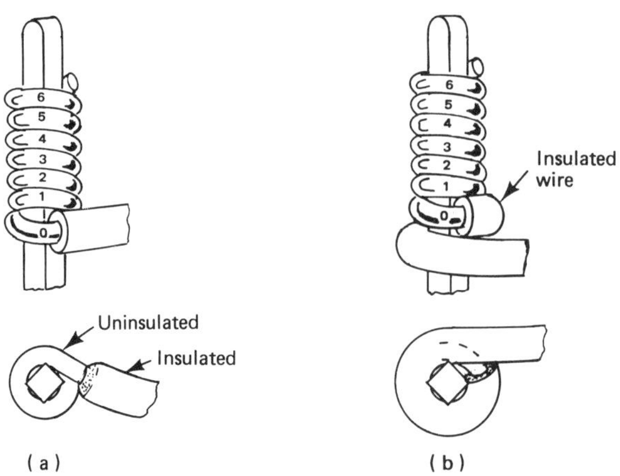

Figure 6–4 Classes of wrap. (a) Standard wrap. (b) Modified wrap.

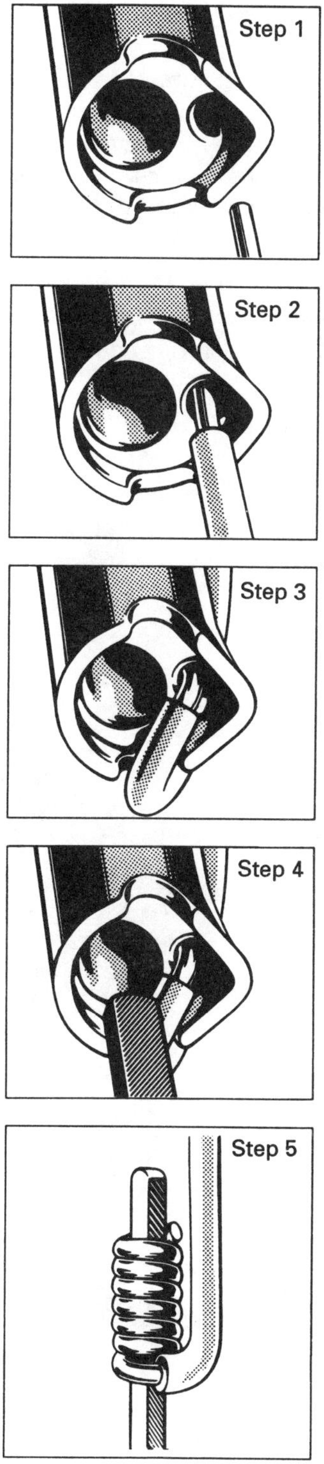

Figure 6–5 Sequence for forming a proper wire wrap.

terminal post, the oxide layer on the wire and terminal is stripped off the surface, forming a good metal to metal connection. An acceptable wrap has five to eight turns, depending on the wire gauge and type (class) of wrap. Appendix D shows a table for the proper number of wraps for class A and class B wraps (Fig. 6–5). The sequence for forming the wrap is as follows:

1. Prepare the wire (strip) for wrapping.
2. Insert the wire into the feed slot in the rotating bit.
3. Anchor the wire around the anchoring notch.
4. Insert the wire onto the terminal post.
5. Finish the connection (class 1).

(a)

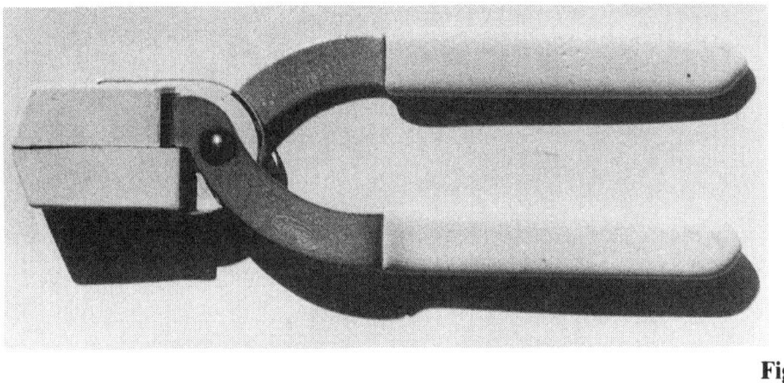

(b)

Figure 6–6 Wire-wrap wire strippers. (a) Manual wire strippers for wire wrap. (Courtesy of Cooper Tools, Inc.) (b) No nick strippers.

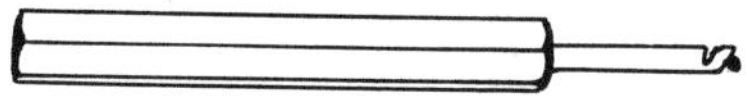

Figure 6–7 Dewrap tool.

6.6.2 Wrapping Rules

When using either automatic hand-held tools or manual tools, a few basic rules must be followed:

- Remove the insulation from the wire without *nicking* the wire. Use a manual (Fig. 6–6) or automatic stripper.
- Do not use thermal strippers as they tend to burn the insulation.
- Position the wire so that it does not pull (unwind) from its connection point when making the next connection.
- When dressing the wires, use only nonmetallic, blunt instruments.
- A dewrapping tool should be used to remove a wire (Fig. 6–7).
- Select the proper bit and sleeve size for the wire gauge being used (see Appendix E).

6.7 WIRE-WRAP LIST

The first step in the process of wire wrapping is for a drafter or technician to make a detailed wire list. A wire list is a table or tabulation that indicates the origin and destination of each wire connection on a board. In other words, it is a listing of where each wire starts and ends. Each line is called a *run*. Figure 6–8 shows part of a schematic diagram and its accompanying wire list.

U16-1	to	U13-2	to	U10-3
U16-2	to	U18-5		
U16-3	to	U13-9	to	U10-4
U16-4	to	U14-6		
U16-5	to	+15V		
U16-6	to	U18-9		
U16-7	to	U14-8		
U16-8	to	U13-5	to	U10-5
U16-9	to	U18-12		
U16-10	to	U13-2	to	U10-6
U16-11	to	U14-11		
U16-12	to	gnd.		
U16-13	to	U17-14		
U16-14	to	——		
U16-15	to	U18-2		
U16-16	to	U14-3		

U18-1	to	U18-4	to	U18-10	to	U18-13	to . . .
U18-2*							
U18-3	to	U13-11					
U18-6	to	U13-8					
U18-8	to	U13-6					
U18-11	to	U13-3, -etc.					

*This connection is not given here since it was listed in the run with C16-15.

Originally, assemblers worked directly from a wire list. Considering that the list in Fig. 6–8 is approximately 1/1000th of a typical wire list, it is easy to see that hand wrapping would be a slow and tedious task, resulting in frequent errors. Therefore, a faster and more accurate method of wrapping is required.

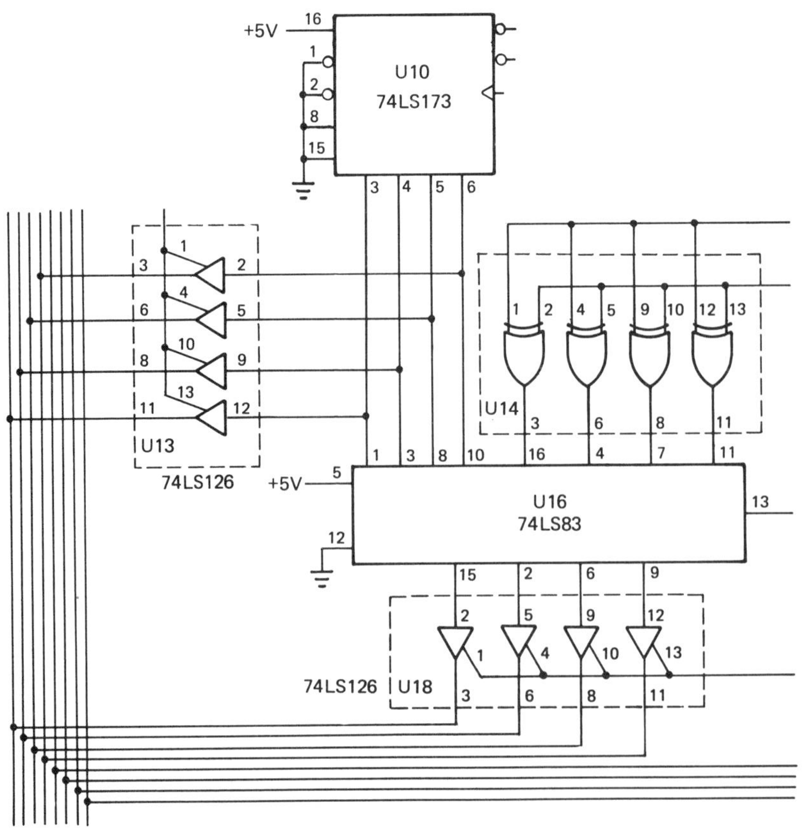

Figure 6–8 Partial schematic drawing and accompanying wire list.

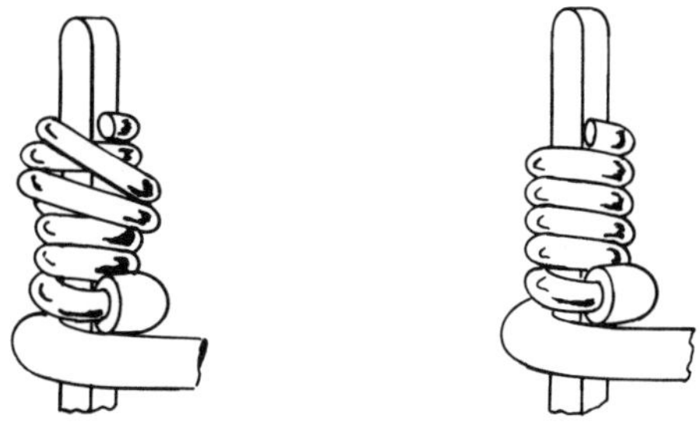

(a) (b)

Figure 6–9 (a) Overwrap and (b) acceptable wrap.

6.8 ACCEPTANCE AND REJECTION CRITERIA FOR WIRE WRAP

To perform successful wire wrap, use the following procedures:

- Let the wire-wrap tool do the work. Do not press too hard because excess pressure can lead to overwrapping. Overwrap may also occur when the wrapping bit has not been selected to correctly match wire gauge and terminal dimension (Fig. 6–9).
- Keeping the tool on the terminal until the wrap is complete is important. If removed too soon, spiral and open wraps may result (Fig. 6–10).
- Feeding wire into the slot in the bit correctly is not difficult. The stripped end of the wire should be pushed in all the way. Not doing so could result in insufficient turns (Fig. 6–11).
- The wire wrap connection should be positioned so that it does not extend onto the tapered section of the terminal (Fig. 6–12).
- There should be no overlapping of adjacent wraps (Fig. 6–13).
- There should be no bare wire not wrapped (Fig. 6–14).
- Bare jumper wire is permissible only when used to connect adjacent terminals (Fig. 6–15).

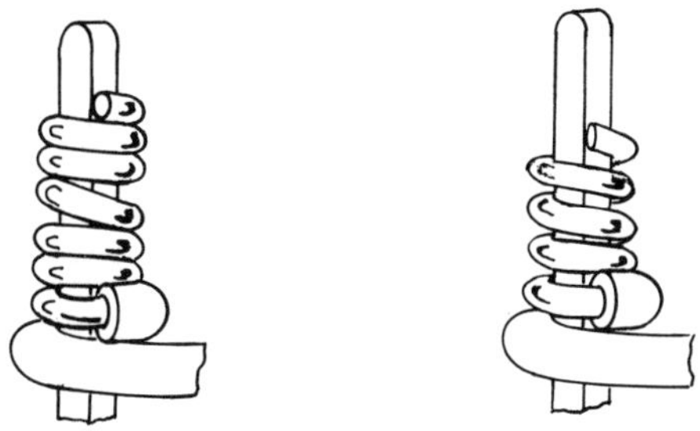

(a) (b)

Figure 6–10 (a) Open wrap and (b) spiral wrap.

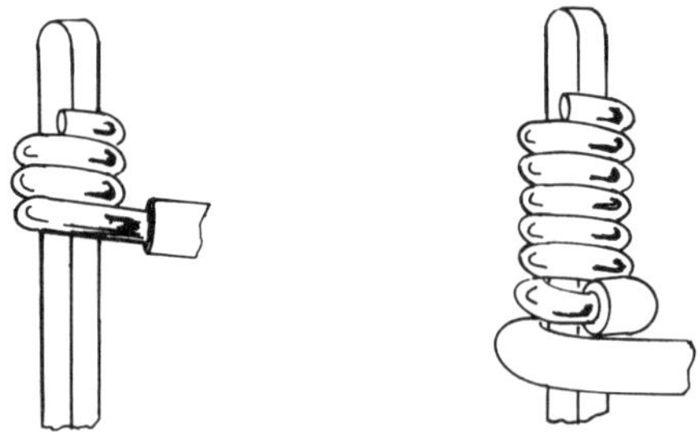

(a) (b)

Figure 6-11 (a) Insufficient turns and (b) sufficient turns.

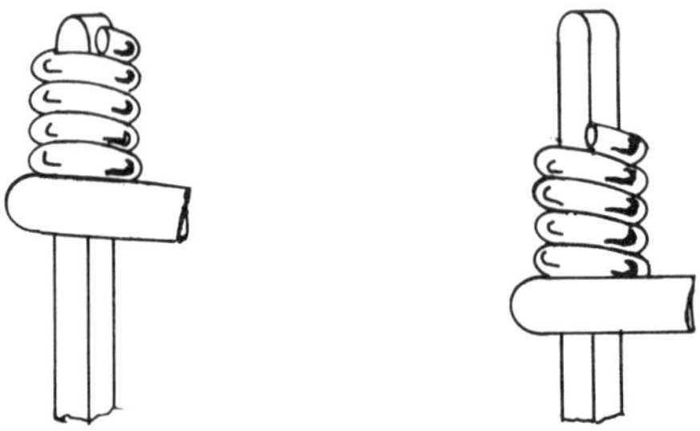

(a) (b)

Figure 6-12 (a) Improperly positioned and (b) properly positioned.

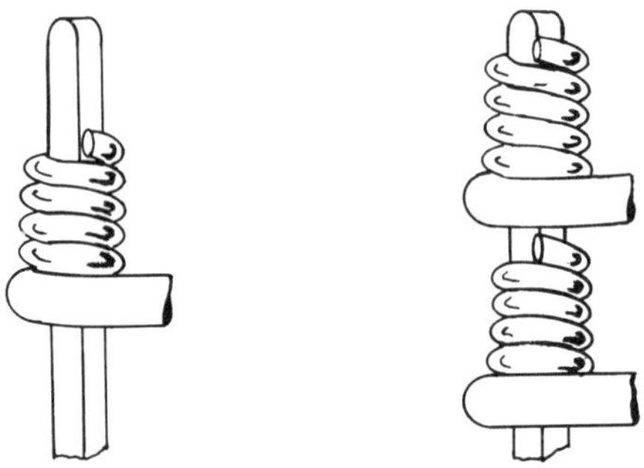

(a) (b)

Figure 6-13 (a) Overlapping and (b) no overlapping.

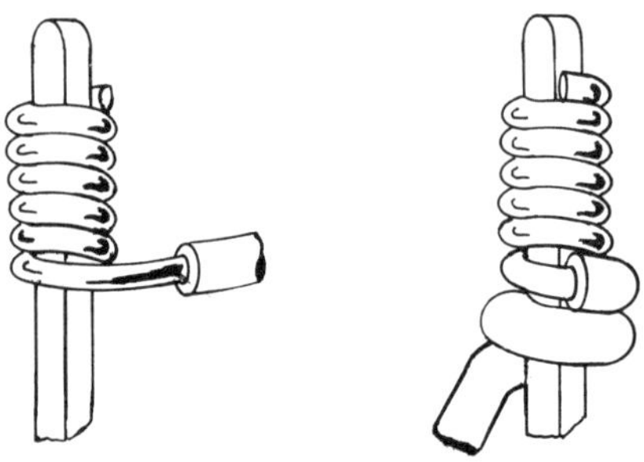

(a) (b)

Figure 6-14 (a) Bare wire is not wrapped and (b) pigtail of the bare wire is flush against the terminal.

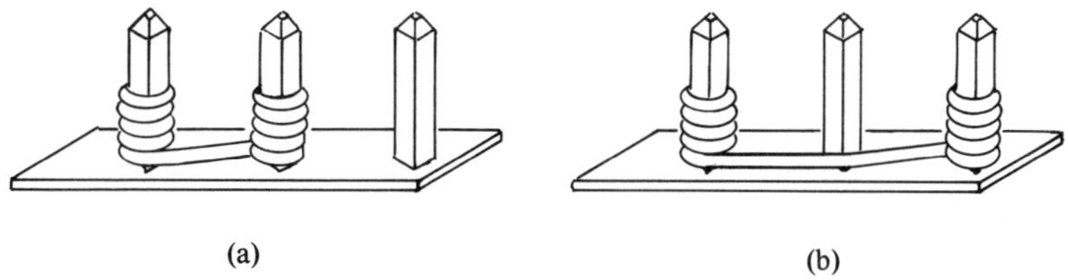

Figure 6–15 (a)Acceptable use of jumper wire and (b) unacceptable bare jumper wire used between nonadjacent terminal posts.

- Wire wrapping is a precision technique. Proper bit and sleeve selection is important. An incorrect choice can lead to problems ranging from pigtails to loose wraps (Fig. 6–16).
- The terminal should be free of cracks and should not be bent in excess of 15° or twisted more than 20° (Fig. 6–17).

6.9 SEMIAUTOMATIC WIRE-WRAP PROCESSING

Semiautomatic systems are computer controlled. A wire list is entered into a computer, which translates the data into a code to automatically move the wire-wrap machine (spindle) to the proper location. Semiautomatic systems can make approximately 300 wraps per hour. Wires are precut to various lengths and then automatically stripped. The operator does the following:

- Inserts the correct length of wire.
- Triggers the spindle to wrap the wire around the post.
- Inspects the connection.
- Instructs the computer to advance to the next location.
- If the connection is an improper wrap, dewraps the connection.

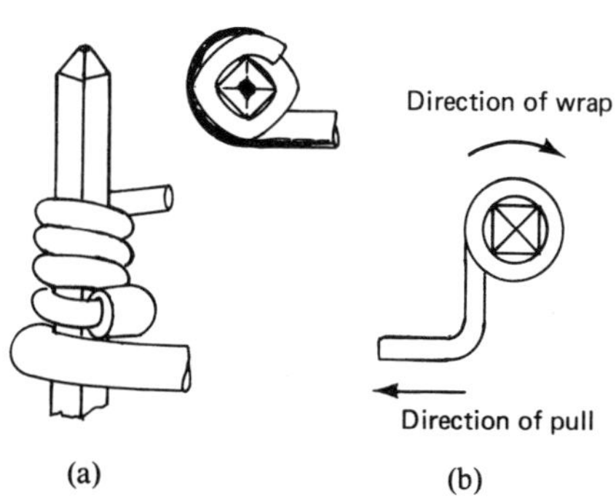

Figure 6–16 (a) Pigtail. (b) Loose wrap tension on wire will unwrap connection.

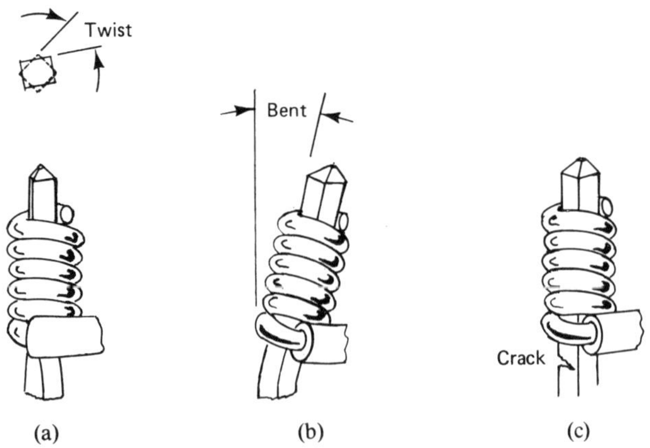

Figure 6–17 Terminal twisted. (a) Terminal twisted more than 20°. (b) Terminal is bent more than 15°. (c) Terminal is cracked.

- Resets the spindle to repeat the wrap or wraps in the run.
- During the process, views the data or information on the computer monitor.

6.10 AUTOMATIC WIRE-WRAP SYSTEMS

Automatic wire-wrap systems are also computer controlled. They do not require a full-time operator, therefore allowing one operator to control several boards at once. Some systems automatically determine the best route for wires. This eliminates tangled wires and simplifies visual inspections (Fig. 6–18).

Automatic systems accept wire lists from CAD systems or microcomputers using a schematic drawing package that creates a wire list. This speeds up the process time, since there is no need to key in the wire list manually.

Because some systems have large RAM memories, one wire run can be completed on several boards, one after another, without reprogramming, or a single board can be completed before continuing to the next. Automated systems can generate approximately 800 wraps per hours.

WIRE-WRAP TEST

1. Define wire wrapping.
2. Why is wire wrapping used instead of PC boards?
3. What is the name of the tool used for removing wire-wrap connections?
4. Why is it better to use an electric hand-held wire-wrap tool rather than a battery-operated tool?
5. Why is it necessary to use blunt, nonmetallic instruments for dressing wires?

Figure 6–18 Fully automated wire-wrap board. (Courtesy of dataCon Interconnect Sub-systems)

6. Is unwrapped bare wire left for flexibility? Why or why not?
7. What materials are used in wire wrapping?
8. Can there be more than three adjacent wraps on a terminal post? Why or why not?
9. What is the correct number of acceptable wraps for a wire-wrap connection?
10. What is the difference between a standard and a modified wrap? Which is preferred and why?

LAB PROJECT

Materials

- Circuit board with wire-wrap studs or IC sockets with wire-wrap studs
- Wire-wrap tool
- Wire-wrap wire
- Dewrapping tool

Procedure

- Make 20 wire-wrap connections.
- There should be four runs with five connections per run.
- Three of these connections should be made with bare jumper (adjacent connections) wire.
- Dewrap two connections from each run.

Chapter 7

Wire Harnessing

7.1 INTRODUCTION

The assembly of electronic systems very often requires the connection of conductors (wires) between the various components in the system (Fig. 7–1a). In such a system, groups of conductors are routed beside each other in a bundle that is held together by some type of binding. Each group of wires is called a **wire harness** (Fig. 7–1b). A harness differs from a cable in that a harness has *break outs* (branches of one or more wires), while a cable goes from connector to connector or contact point in one continuous length (Fig. 7–1c).

The use of a wiring harness, in contrast to custom wiring, simplifies the wiring of more than one of the same type of equipment, saves money, is neater, and reduces error. A harness is often used when several units are constructed. The first harness is formed and tied in the chassis. But it is not soldered into its permanent position. The harness is then removed to make a *template for a harness board* (Fig. 7–2). The remaining harnesses are produced on the harness board. Custom wiring is used when only one unit is to be constructed. Each wire is located in place, soldered, and tied.

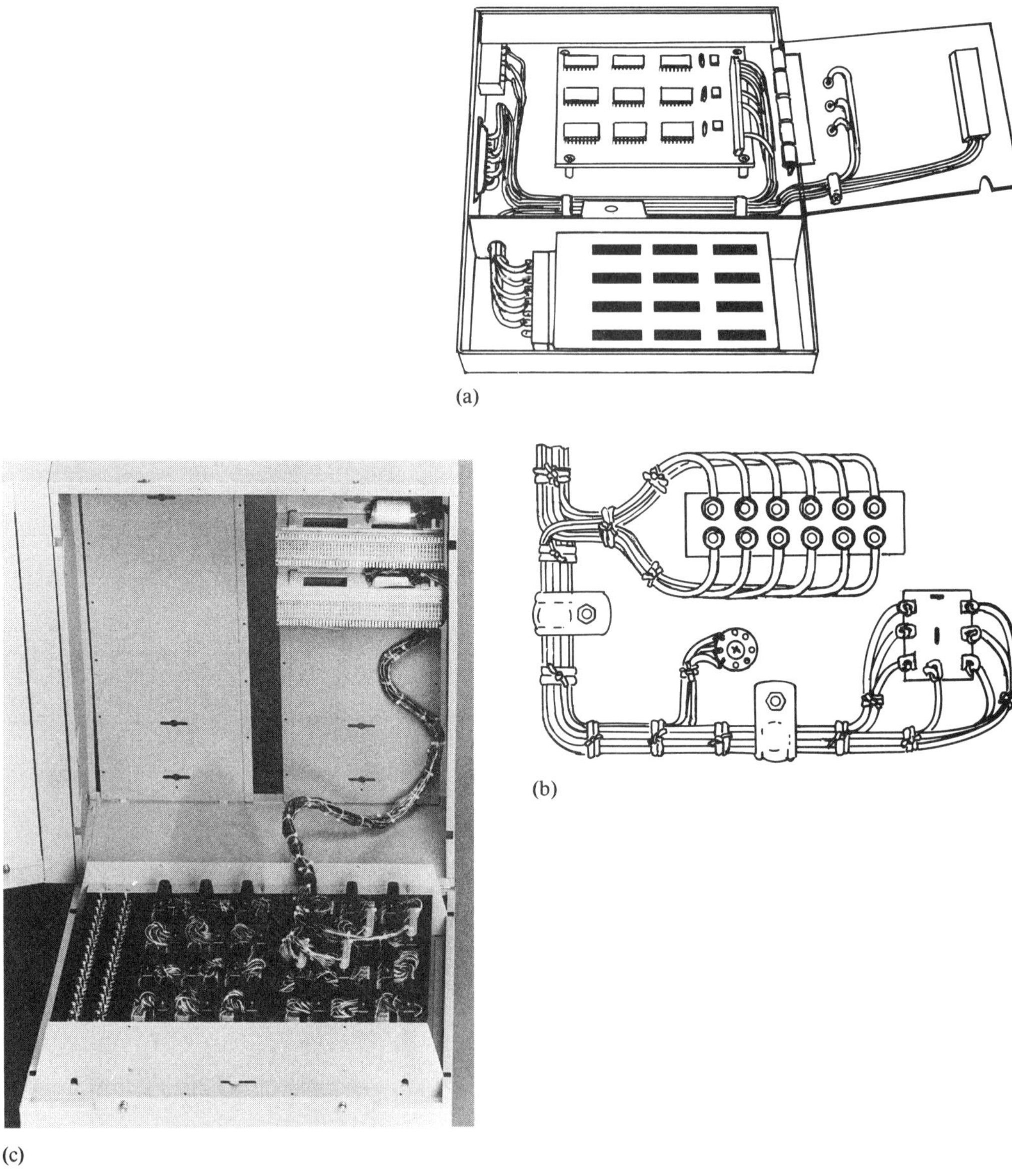

(a)

(b)

(c)

Figure 7–1 (a) Components and devices are connected together by a wire harness. (b) Connecting wires are bound together with lacing twine and cable clamps. (c) Cable run between panels.

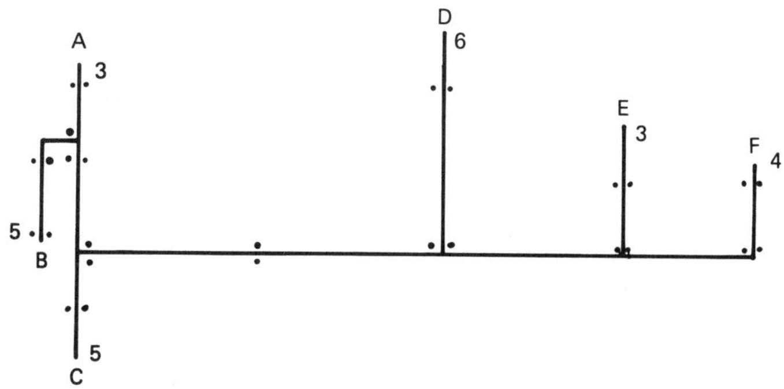

Figure 7–2 Template for a harness board.

7.2 GENERAL HARNESS CONSTRUCTION

In the past the most common method of forming wiring into a harness was lacing. Today the use of plastic tie wraps, zipper tubing, shrink tubing, cable clamps, spiral wrap, and others, has somewhat simplified the holding together of wires, but these items are much more expensive than lacing twine. When cost is a factor, lacing is usually the preferred method.

A harness is carefully formed so that, when installed, the free end of each wire fits neatly into the correct terminal with extra wire to form a ***service loop.*** A service loop is a length of extra wire that is left on each conductor, as shown in Fig. 7–3. This allows the wire to be disconnected and resoldered if a component must be replaced. The ends of the wires in a harness may be fitted with lugs, solder terminals, or connectors.

Wires for a harness are precut and prepared (stripped and tinned) prior to mounting on the harness board. If required, terminals or solder lugs are installed at this time. Color coding or wire labeling is often used to simplify the construction of the harness and the tracing of wires in case of error in assembly. Wire color coding will be explained later in this chapter.

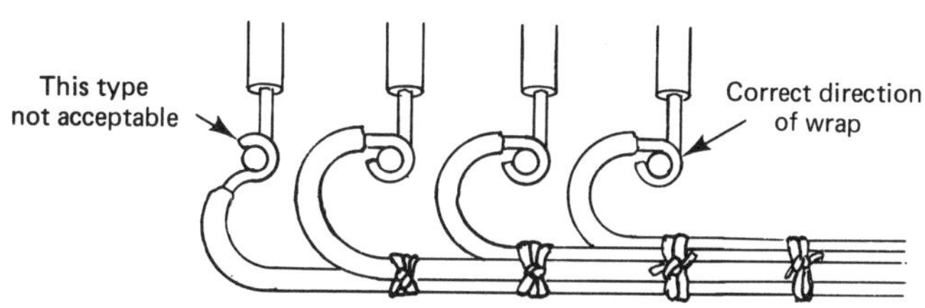

Figure 7–3 Wire harness with a service loop.

7.3 LACING THE HARNESS

Any type of binding, such as zipper tubing, nylon lacing twine, or plastic tie wraps, that is used to hold the wires of a harness in a neat bundle may be called wire lacing. Most lacing twine or tape is impregnated with wax, which prevents slippage of the ties and allows them to be knotted tightly. They must not be tied so tightly that the insulation is compressed. The proper width of nylon lacing tape should be used with each cable. The proper widths are as follows:

- $\frac{1}{32}$ inch for up to $\frac{1}{2}$-inch cables.
- $\frac{1}{16}$ inch for $\frac{1}{2}$- to 1-inch cables.
- $\frac{1}{8}$ inch for 1- to 2-inch cables.

Two common lacing techniques are used in harness making: spot-tie method (Fig. 7–4) and the running-lace method (Fig. 7–5). The running lace is composed of a series of connected loops. When a harness is being constructed, a running lace may be used. When both ends of the wires in a cable are free, the cable is laced from the middle to each end. The starting tie is shown in Fig. 7–5a. Care must be taken to make a snug starting tie.

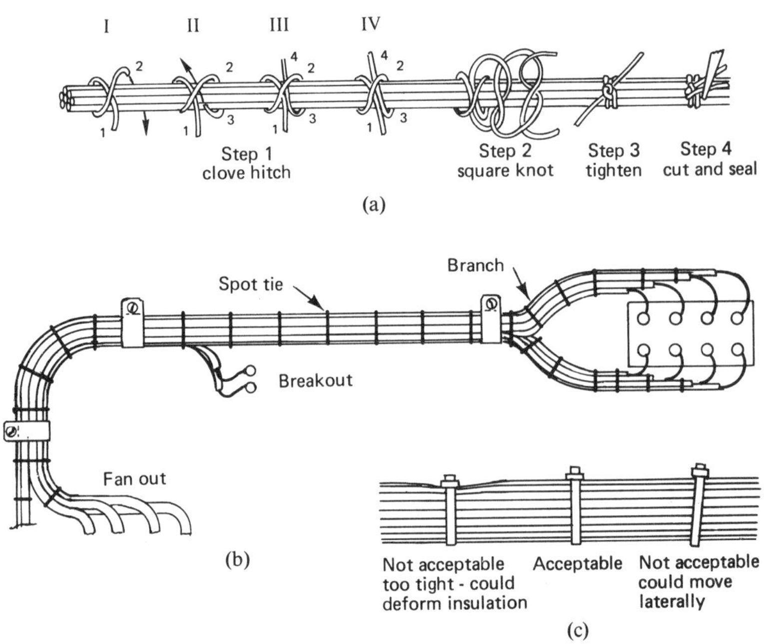

Figure 7–4 Spot ties. (a) Making the tie. (b) Sample of a cable run with spot tying. (c) Acceptable and unacceptable ties.

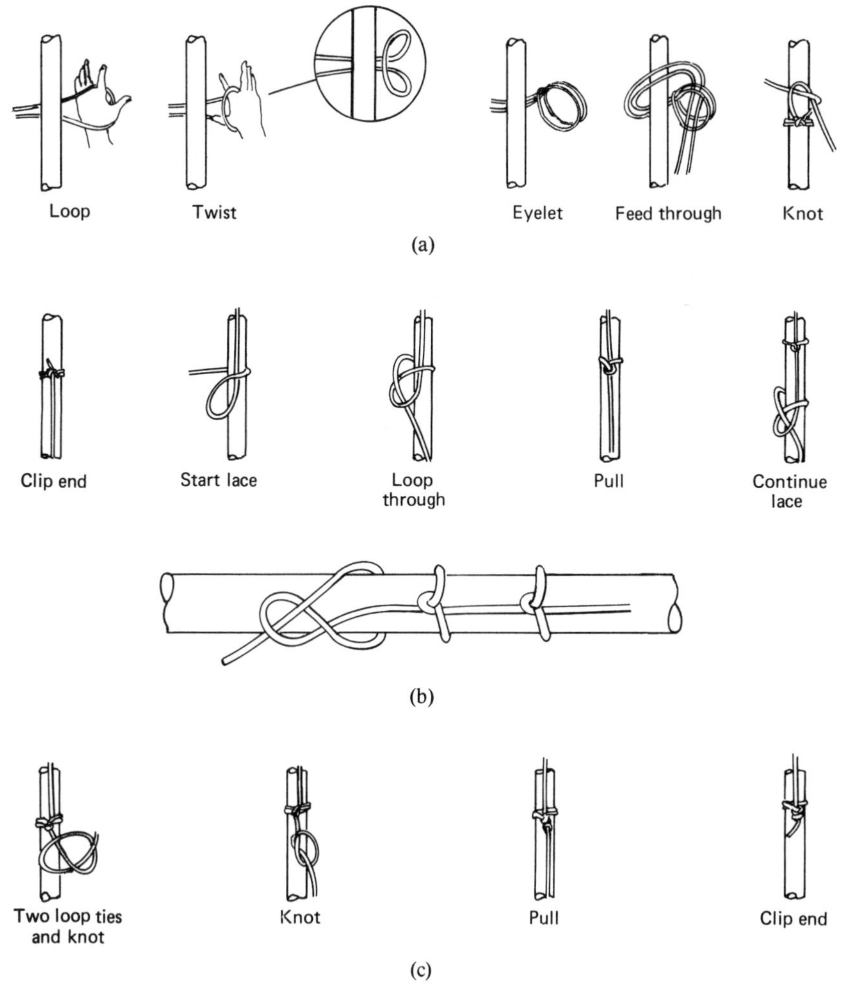

Figure 7–5 Running-lace method. (a) Starting tie. (b) Running lace. (c) Final tie.

Whether using the spot-tie method or the running lace, ties should be evenly spaced. The general rule is that the ties should be spaced no less than $\frac{1}{2}$ inch apart or twice the cable diameter if it is larger than $\frac{1}{2}$ inch in diameter (Table 7–1).

Lacing ties placed at a harness branch are called a lacing breakout. Note that a starting knot is tied at the main trunk and a square knot is tied at the branch, as close to the trunk as possible. Special tying techniques are used at *breakout points* on a harness. Figure 7–6 shows two methods that may be used.

Specific electronic units may require one method of lacing over another or in some cases may exclude lacing altogether. Generally, spot-tie lacing is preferred to the running tie. This is because repairs can be done easier and faster by removal and

Table 7–1 Proper Tie Spacing

Bundle diameter (in.)	Tie strap spacing (in.)
Less than $\frac{3}{4}$	2
$\frac{3}{4}$ to 1	3
1 to $1\frac{1}{4}$	4
$1\frac{1}{4}$ to $1\frac{1}{2}$	5
$1\frac{1}{2}$ up	6

replacement of spot ties, rather than the running lace. The spot-tie method that is preferred today is the mechanical spot tie (tie wraps). Although more expensive than lacing, it is faster and much easier to use (Fig. 7–7).

7.4 DEVELOPING THE HARNESS BOARD

Making a harness board requires a specific sequence of steps. To develop accurate production harnessing:

- If a real unit is not available, construct a mockup of the device, placing connectors, terminals, and the like, at their appropriate locations (Fig. 7–8).
- Attach (do not solder) all wires (conductors) to their contact points.
- Lace the wires into a harness. Make sure all breakouts have been properly tied.
- Recheck conductor length, routing, dress, and so on.
- Remove the harness from the board or mockup.
- Make a one to one harness drawing and attach it to a board, using rubber cement (Fig. 7–9).

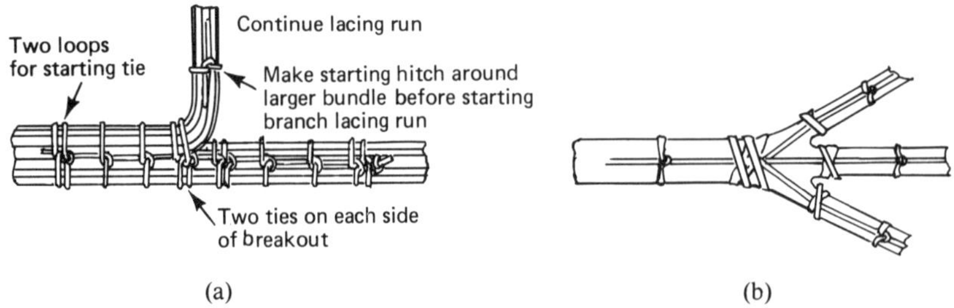

Figure 7–6 Cable tying for breakout points. (a) Breakout at a single branch. (b) Breakout at a three-legged branch.

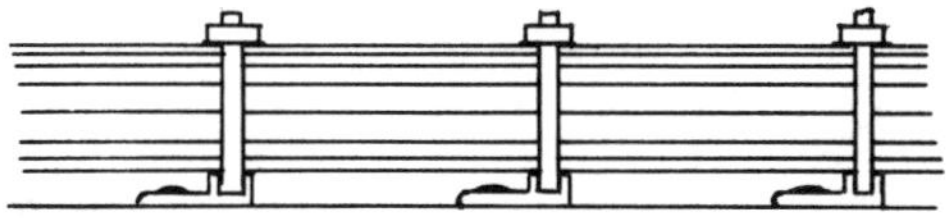

Figure 7–7 Mechanical spot tie.

- Attach pins (small-head finishing nails) or pegs to serve as guides for forming the harness into its proper shape (Fig. 7–10a).
- After completing one harness, and a preferred sequence for wire layout has been established, a wiring list can be developed. An example is shown in Fig. 7–10c.

7.5 WIRE COLOR CODING

Insulated wire used in the electronic industry is generally color coded. The color code is similar to that used with resistors:

0 = black
1 = brown
2 = red

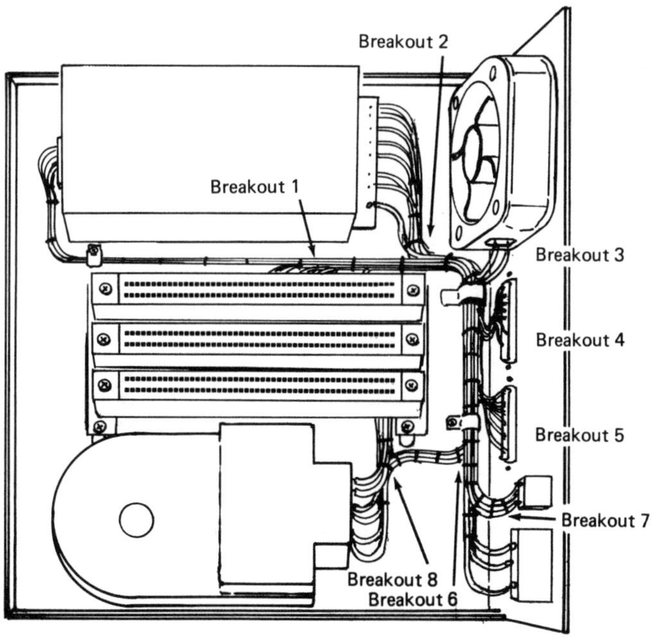

Figure 7–8 Mockup board.

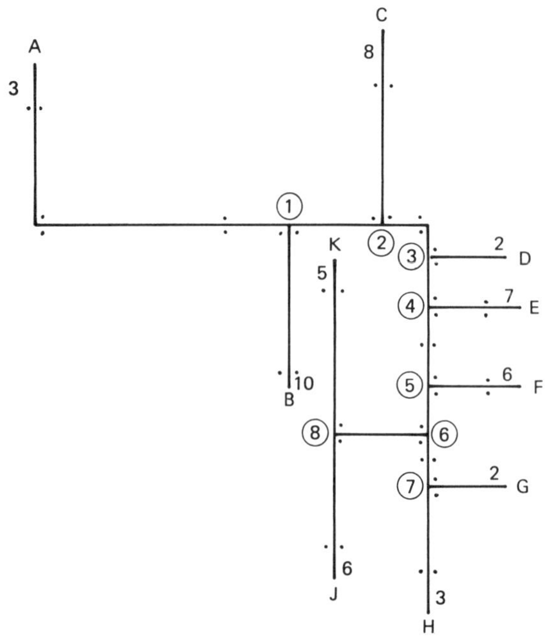

Figure 7–9 Harness drawing with wiring layout.

3 = orange
4 = yellow
5 = green
6 = blue
7 = violet
8 = gray
9 = white

Four possible color code arrangements can be used:

Number of digits in code	Sequence of wire colors
A. 1	Solid-color wire (insulator)
B. 2	First digit, background color. Second digit, color of stripe.
C. 3	First digit, background color. Second digit, wide stripe. Third digit, narrow stripe.
D. 4	First digit, background color. Second digit, wide stripe. Third digit, first narrow stripe. Fourth digit, second narrow stripe.

For example, if the code is 73, the background of the wire is violet and the single stripe is orange. If the code is 652, the background is blue, the wide stripe is green, and the narrow stripe is red (Fig. 7–11).

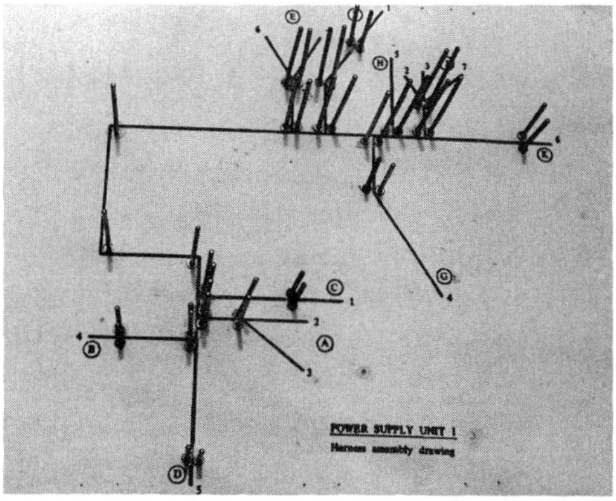

(a)

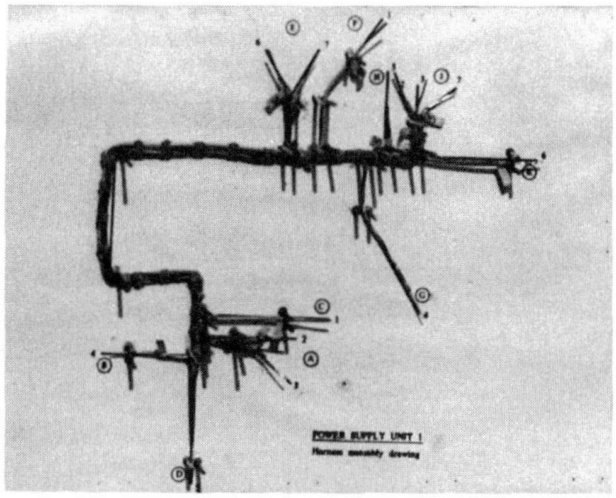

(b)

Wiring Sequence List

Power supply unit 1

Item no	Route	Wire identification no	AWG	Color	Length (inches)
1	C–F	1	18	999	12
2	A–J	2	18	777	12 1/4
3	A–J	3	22	990	12 1/4
4	B–G	4	22	915	13
5	D–H	5	22	000	12 1/4
6	E–K	6	18	222	6
7	E–J	7	22	990	5

(c)

Figure 7–10 Harnessing. (a) Board with pegs in place. (b) Board with harness in place. (c) Wiring sequence list.

7.6 ALTERNATIVE LACING METHODS

There are three common methods of holding wires together in a bundle without lacing twine: tie wraps, sleeving, and spiral wrap. Spiral wrap, a narrow strip of plastic formed into a spiral, is a simple method for holding wires in a bundle (Fig. 7–12a). To lace the cable, the spiral is twisted around the wires to form a neat bundle. The spiral can be precut to the desired length.

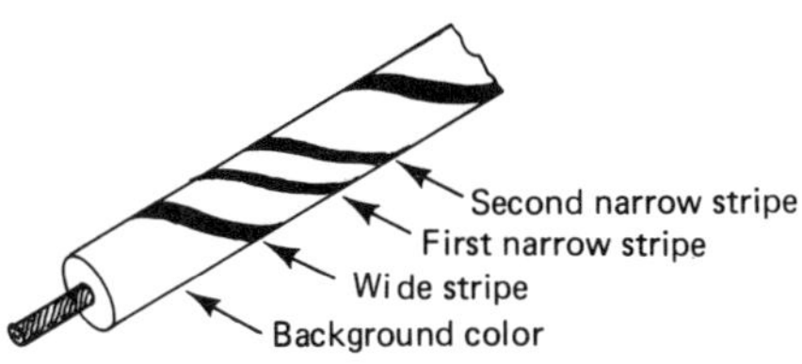

Figure 7–11 Wire color code.

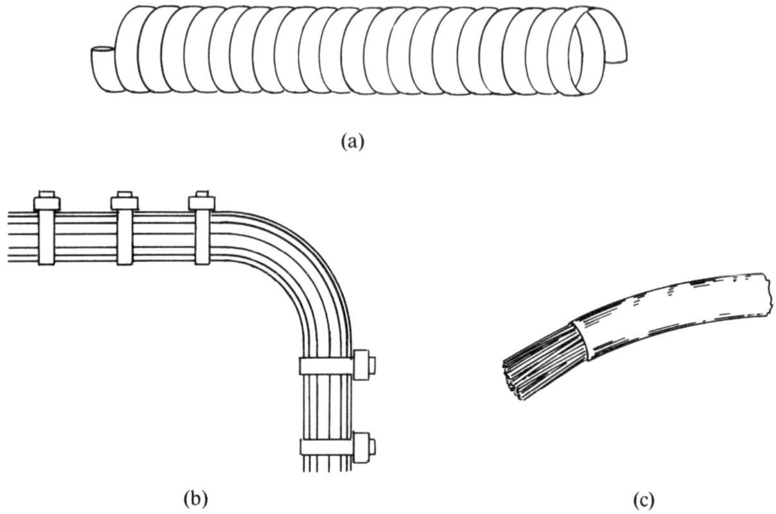

Figure 7-12 Alternative lacing methods. (a) Spiral wrap. (b) Tie wrap. (c) Sleeving.

When using sleeving, a cable is formed by placing the wires in a sleeve (Fig. 4-12c). Sleeving is also used to cover splices and component leads to prevent the bare wires from shorting with other components and wires.

Three common types of sleeving (tubing) are used: zipper tubing, shrink tubing, and loose tubing. Shrink tubing is commonly used to cover wire splices and solderless connectors. The method for its use will be explained in Chapter 8. Loose tubing is generally used to form cables. It's also used to cover splices and soldered and solderless connector and component leads.

To form a cable with loose tubing:

- Cut the tubing to the desired length.
- Tie the end of the bundle of wires with lacing twine.
- Thread the end of the lacing twine through the tubing.
- Pull the bundle of wires through the tubing.
- Secure the tubing around the wire by using a spot tie (Fig. 7-13).

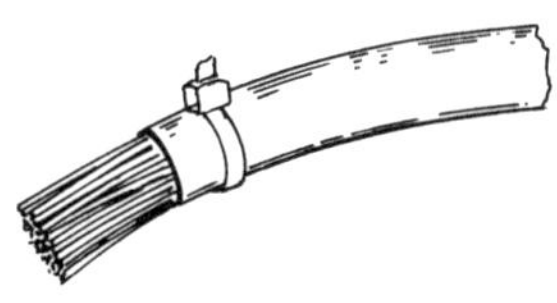

Figure 7-13 Securing a cable or harness using loose tubing.

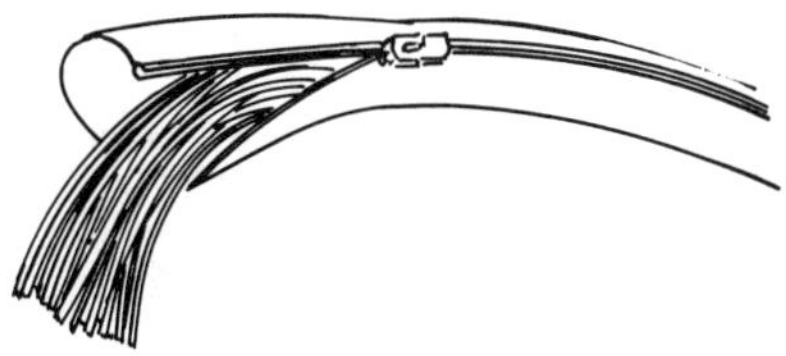

Figure 7-14 Using zipper tubing to form a harness or cable.

Zipper tubing is frequently used instead of loose tubing because it is easier to install. However, zipper tubing is more expensive and does not come in the variety of sizes that loose tubing does (Fig. 7-14). Zipper tubing is more appropriate for large wire bundles.

Tie wraps or straps are made of plastic or nylon. They are used in place of spot tying to form a wire harness or cable. Most tie wraps have self-locking clips that lock the tie in place once it has been pulled tight (Fig. 7-15).

7.7 SECURING A HARNESS

Once a harness is formed and soldered, it is necessary to *anchor* it in place to protect it from hot or moving parts, sharp edges, and damage from vibration. Metal or plastic clamps can be used to secure or anchor it in place (Fig. 7-16).

When forming and securing a wire harness or cable, the following criteria should be considered:

- Harnesses shall be routed according to the drawing or harness board. No alterations will be made in the routing.

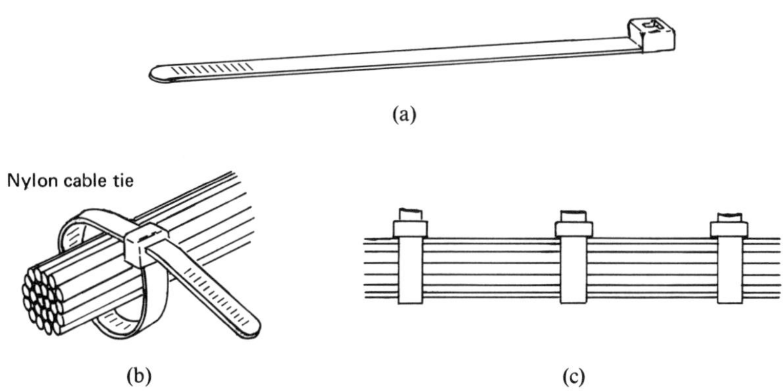

Figure 7-15 Using nylon tie wrap. (a) Nylon tie wrap.(b) Looping a tie wrap around a wire cable or harness.(c) Finished product.

Table 7-2 Bend Radii

Conductor	Minimum bend radius
Insulated wire	2 × outside diameter of wire
Shielded insulated wire	4 × outside diameter of wire
Coaxial and multiconductor jacketed cable	6 × outside diameter of wire
	(5 × outside diameter for polyethylene)
Harnesses	Minimum allowable for single conductor in cable

- Wire bundles shall be secured by the method specified on the drawing (tie wraps, spot ties, and so on).
- The minimum bend radius for any part of the harness, including breakouts, shall be per Table 7-2. There should be no broken or *stressed* wires (Fig. 7-17).
- Wires shall not loop in and out of lacing, and all conductors shall lay straight and parallel (Fig. 7-18).
- Cable clamps should be installed as specified on the drawing. They shall be tightly secured to prevent movement of the harness within the clamp. No wires in the bundle should be damaged or pinched by the clamp (Fig. 7-19).
- When secured by clamps, there shall be minimal sagging or flexing of the harness or the breakout leads (Fig. 7-20).
- Standard cable clamp stackups should be as shown in Fig. 7-21, unless drawing specifies differently.

WIRE HARNESSING TEST

Two choices are given for each question. Circle the correct choice.

1. Groups of wires form a ________ (cable *or* mold).
2. Each wire in a harness is formed with extra wire to form a ________ (conductor *or* service loop).
3. Harnesses are held together by ________ (ties *or* tape).

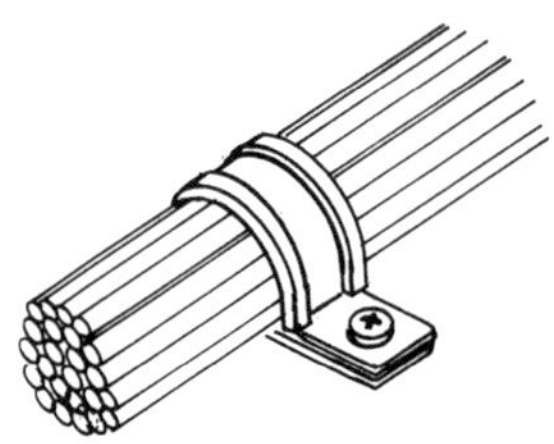

Figure 7-16 Securing a wire harness with cable clamp.

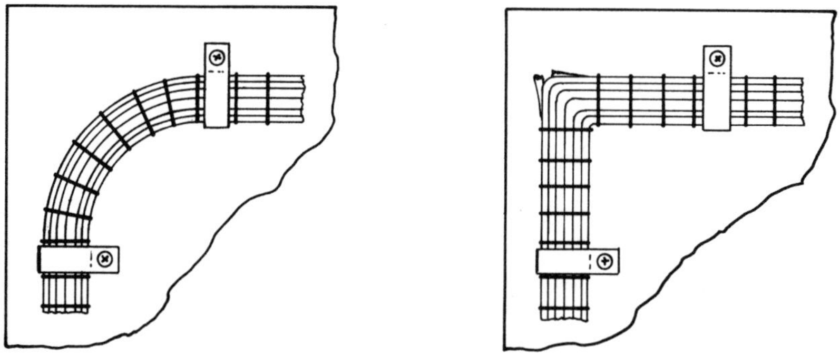

(a) (b)

Figure 7–17 Proper bend radius. (a) Acceptable: no broken or strained wires; bend radius meets minimum requirements. (b) Unacceptable: sharp bend and broken wire; bend radius less than minimum requirements.

(a) (b)

Figure 7–18 Harnessing and harness support. (a) Acceptable: wires are straight and parallel with no looping in and out of lacing. (b) Unacceptable: wires loop in and out of lacing.

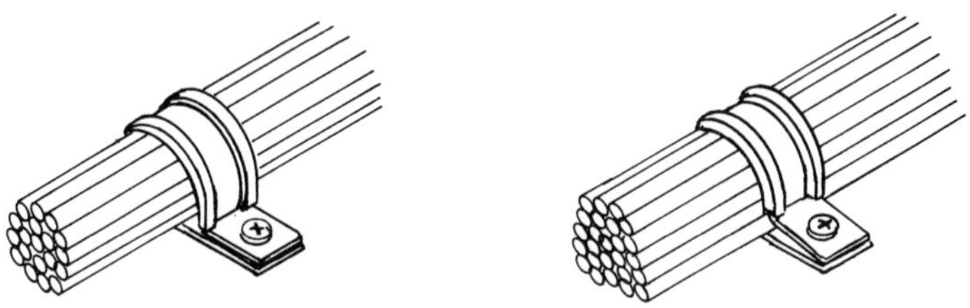

(a) (b)

Figure 7–19 Securing a harness. (a) Acceptable: no damage of wire bundle. (b) Unacceptable: wire is pinched in the clamp.

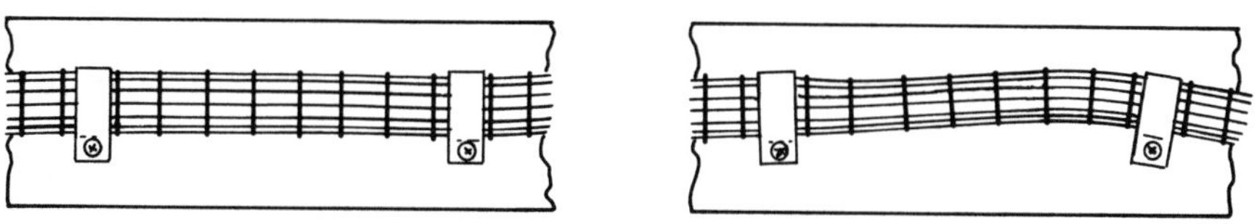

(a) (b)

Figure 7–20 Supporting a harness. (a) Acceptable: harness is secured properly by clamps; minimal sagging. (b) Unacceptable: harness sags.

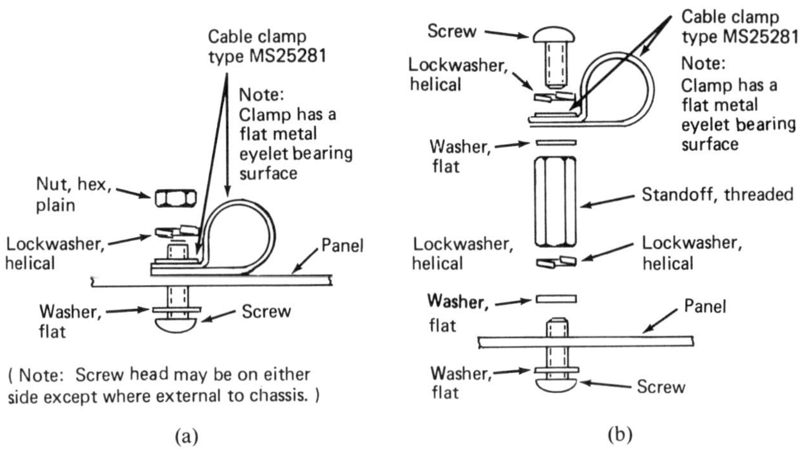

Figure 7-21 Proper clamping. (a) Clamped directly to chassis. (b) Clamped using a standoff.

4. Color coding or labeling of wires allows for easier ________ (soldering *or* trouble-shooting).
5. Where should you use a lacing breakout? (at a harness bend *or* at a harness branch)
6. The nylon twine used to tie a cable assembly is called ________ (wax thread *or* lacing twine).
7. Where is the starting tie secured at a branch breakout? (to the branch *or* to the trunk)
8. Cable lacing should be tied tight enough to ________ (prevent movement *or* distort the insulation on the wire).
9. Which cable tie should be used to simplify repair or replacement of wires in a harness? (spot tie *or* running wire)
10. How is a cable laced when both ends are free? (from the middle to each end *or* from end to end)
11. The two lacing techniques commonly used in forming a harness are ________ (running lace and square knot *or* running lace and spot tying).
12. A pattern used for forming similar items is called a (guide *or* template).
13. What is the purpose of sleeving? (hold wire together without lacing *or* maintain the heat within the wire)
14. Two types of sleeving are (zipper and shrink tubing *or* shrink tubing and lacing twine).
15. Cable clamps are used to prevent ________ (removal of the harness *or* movement of the harness).

LAB PRACTICE: MAKING A HARNESS

Materials

- Harness board
- Wires cut to proper length
- Lacing twine and tie wraps

Procedure

- Using a wiring sequence list and wires supplied, assemble the harness on the harness board.
- Use spot ties to form the harness, using nylon tie wraps to form three of the ties.
- Check your ties to be sure they are spaced according to the specifications table.
- Check to make sure the ties are firmly secured to prevent movement, but not tight enough to crimp the insulation.

Chapter 8

Wire Splicing and Electrical Connections

8.1 INTRODUCTION

Electrical connections must be reliable in order for a system to function properly. So far we have discussed making connections by soldering to PC boards, terminals, and wire wrap. Two other methods for making reliable electrical connections are **splicing** and **solderless connections**. Solderless or *crimp*-type terminals (Fig. 8–1a) are sometimes used as a means to provide a strong mechanical connection (metal to metal). They are quick and easy to make and are as durable and electrically conductive as soldered terminals. Splicing is a method of connecting two or more wires electrically without using a stationary terminating point (Fig. 8–1b). Splicing is not typically used in electrical wiring, but sometimes it is necessary for repairing a break or to lengthen a conductor.

8.2 CONDUCTORS

Factors that must be considered in electrical circuits besides a good electrical connection are the resistance and current-carrying capability of the conductors. Wires that

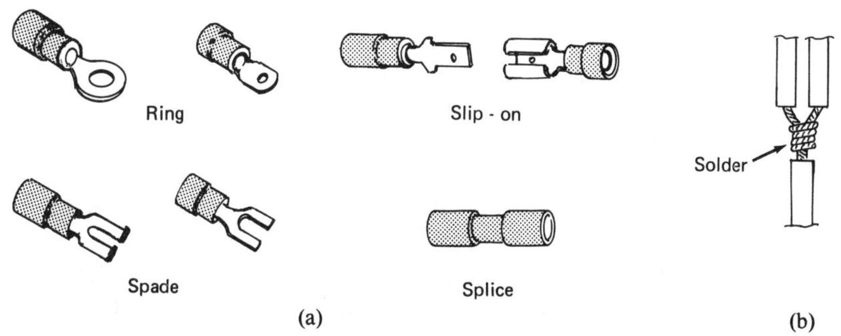

Figure 8-1 Electrical connections. (a) Solderless connectors. (b) Splice.

have a high resistance and thus cannot carry the current needed in the system will be damaged from excessive heat.

Any material that permits the free flow of electrons is classified as a conductor. Various metals, such as aluminum, gold, and silver, are used as conductors. However, copper is the most common because of its low resistance and relatively low cost. Some typical conductors are shown in Fig. 8-2.

We learned earlier that a conductor must be covered with insulation to prevent short circuits between conductors or between a conductor and some metal surface.

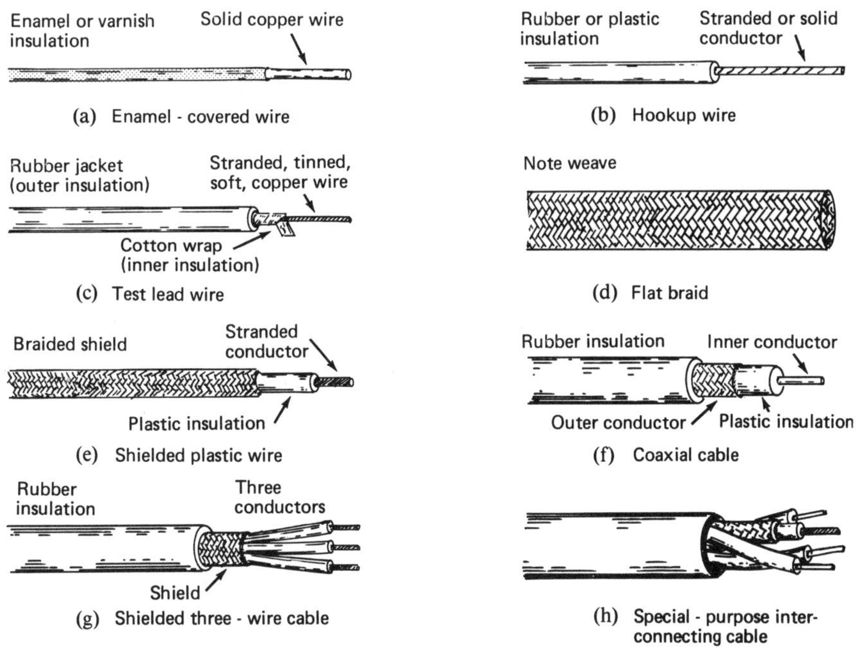

Figure 8-2 Typical conductors.

The important factor in choosing the type of insulation is the breakdown voltage. **Breakdown voltage** is the maximum safe voltage that can be placed on a conductor before the insulation breaks down. Other factors such as heat, resistance, and wear resistance may be important in the selection of insulation for a conductor.

The current-carrying capability of a conductor depends on the cross-sectional areas of the conductor and the material from which the conductor is made. The table in Appendix F provides information on copper conductors, such as the gauge, diameter, current-carrying capacity, and resistance per 1000 feet. For example, suppose we want to know the diameter, current capacity, and resistance of 1000 feet of no. 20 wire. The table in Appendix F shows that no. 20 wire has a diameter of 0.03196 inch. It can handle as much as 1.46 amperes of current and has a resistance of 10.15 ohms/1000 ft. If we needed to know the resistance of 10 feet of no. 20 wire, we would divide 10.15 ohms/1000 feet times 10. The answer would be 0.1015 ohms.

$$(10.15/1000 \text{ ft}) \times 10 = 0.1015 \text{ ohms}$$

Note that the values given in the table are at a temperature of 25°C.

Figure 8–3 shows a wire gauge used to determine the gauge of a conductor. Note that the smaller the wire gauge number is the larger the conductor diameter and the larger the current-carrying capability.

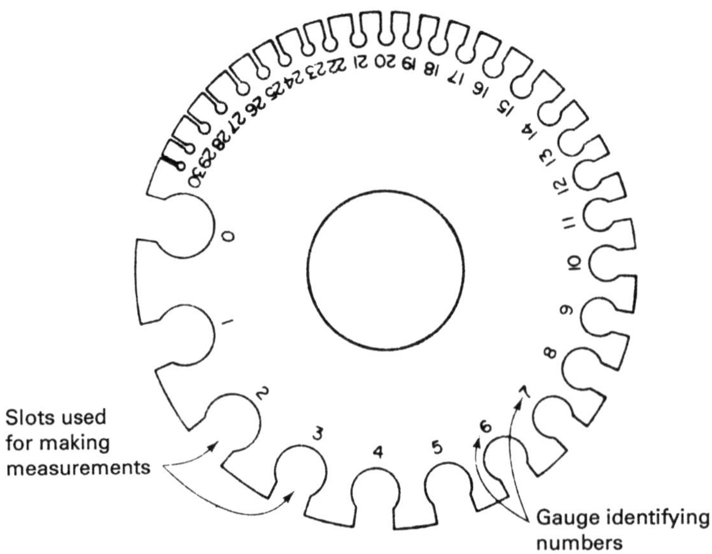

Figure 8–3 Wire gauge used to determine the AWG number of a conductor.

8.3 SPLICING

Splicing is a means of mechanically connecting two or more wires together to form an electrical circuit. The three most common wire splices are the rattail, tee, and Western Union. To ensure a good electrical connection, the conductors must be properly prepared prior to making the connection.

- Remove insulation from the area to be spliced. Some conductors use enamel as an insulation. To remove this type of coating, use sandpaper, scrape with a knife, or use *lacquer thinner.*
- Clean the surfaces of the conductors.
- Tin the exposed area of the conductors to be spliced.
- If sleeving is to be used, cut a piece to the desired length, and slip the tubing over one of the wires. This must be performed prior to joining the wires together.

8.3.1 Rattail Splice

Rattail splices are generally used when joining two wires together, such as in an electrical junction box. Twist the two conductors fingertight, approximately five to six turns. If the wire size is larger than AWG 14, pliers should be used for twisting the wires together. Be careful not to damage (nick) the wires with the pliers (Fig. 8–4). Solder the connection and cover with sleeving, shrink tubing, or electrical tape.

If using shrink tubing, slide the tubing into position and heat using a heat gun. The tubing will shrink and seal the bare splice (Fig. 8–5).

Rattail splicing can also be made with solderless connections such as wire nuts or crimp connections, as will be discussed later under the topic of solderless connections.

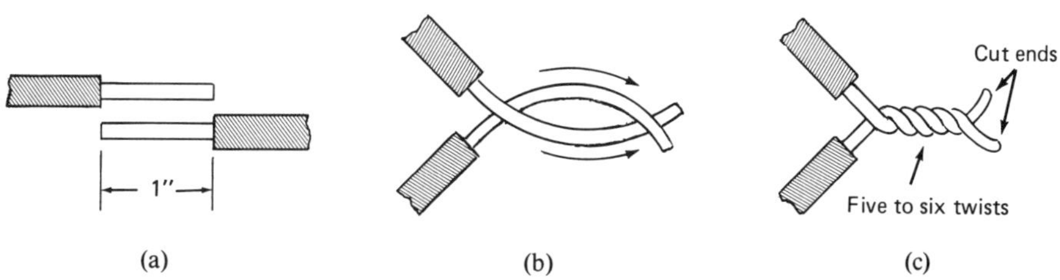

Figure 8–4 Rattail splice.

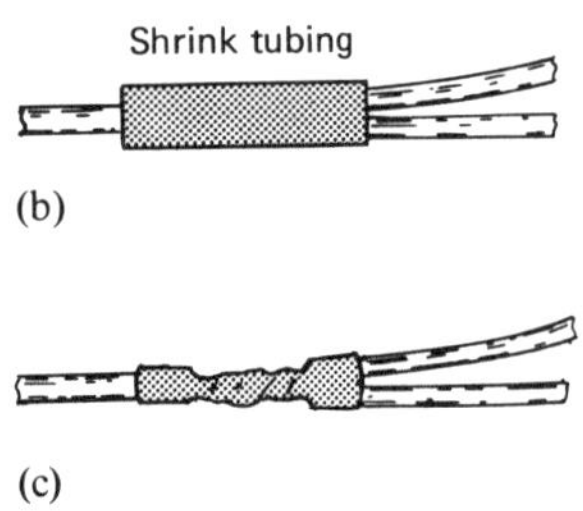

Figure 8–5 Using shrink tubing. (a) Heat gun. (b) Tubing before shrinking. (c) Tubing after shrinking.

8.3.2 Tap or Tee Splice

The **tap** or **tee splice** is generally used for tapping into existing wiring. With this type of splice, the existing wire is not cut, but is stripped where the branch wire is to be joined. Examples of a tee splice are tapping into the wire of the positive voltage in an automobile when adding a stereo system or connecting a conductor to a wire that runs through a junction box in home electrical wiring. To perform a tee splice, strip approximately 0.75 inch of insulation from the main wire. Join the two wires as shown in Fig. 8–6. Solder and wrap with electrical tape. Sleeving cannot be used with this type of splice.

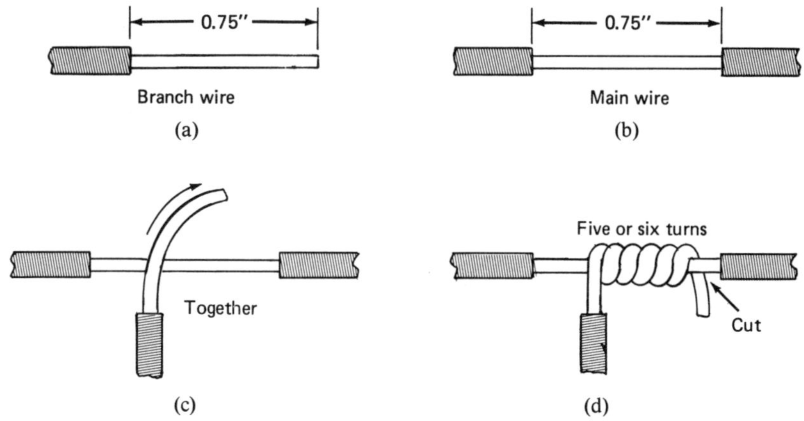

Figure 8–6 Tee splice.

8.3.3 Western Union Splice

The **Western Union splice** is the strongest of the three splices. This type of splice is used for repairing cut or broken wires. The splice is made by crossing the stripped ends of two wires (strip approximately 1 inch) and twisting them together as shown in Fig. 8–7. The wires are then soldered and wrapped with electrical tape, sleeving, or shrink tubing.

8.4 SOLDERLESS CONNECTIONS

A solderless connection is made by joining two or more wires by pressure, without the use of solder. These types of connections are used when it is necessary to remove conductors from their termination point or when there is little chance of vibration of the wire.

A **wire nut** (Fig. 8–8a) is a solderless connector with tapered internal threads that is twisted over the exposed ends of a rat tail splice. When using a wire nut, the splice is not soldered. Care must be taken that the ends of all the wires are even and

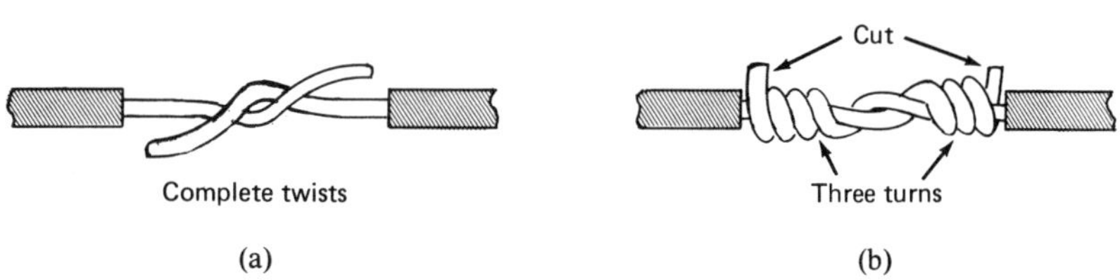

Figure 8–7 Western Union splice.

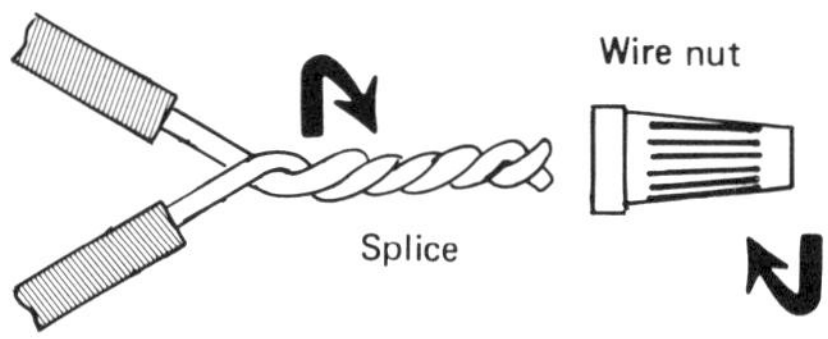

(a)

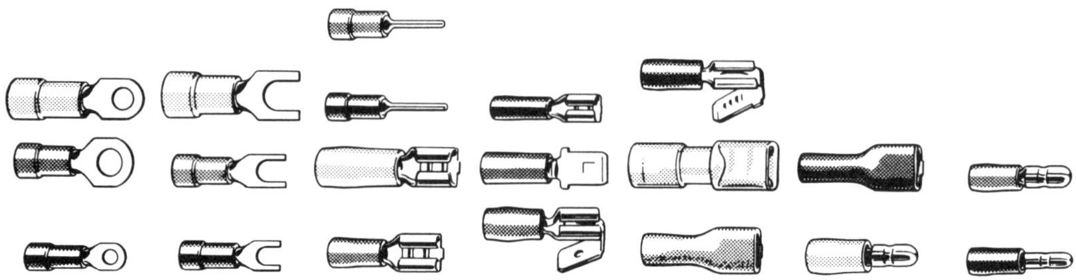

(b)

Figure 8–8 Solderless connectors. (a) Wire nut. (b) Terminal lugs.

enter the wire nut together. Several sizes of wire nuts are available to accommodate various sizes of wire. It is recommended that no more than three wires be secured with a wire nut.

8.4.1 Solderless Terminal Lugs

Examples of solderless terminal lugs are shown in Fig. 8–8b. The connector's back portion has a hollow cylinder to allow it to accept a wire. After the bare wire is inserted into the cylinder, a crimping tool (Fig. 8–9) is used to clamp the metal contact

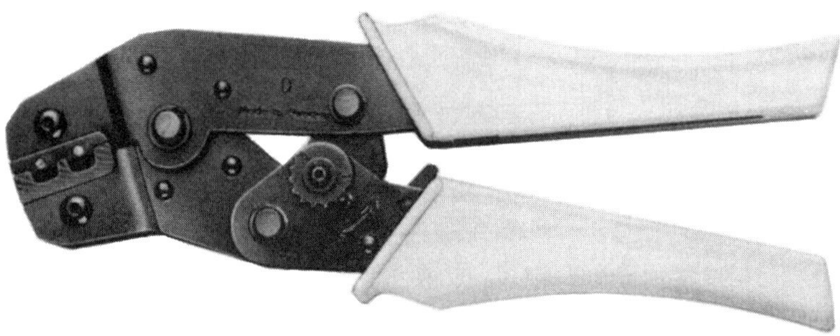

Figure 8–9 Crimping tool. (Courtesy of Cooper Tools, Inc.)

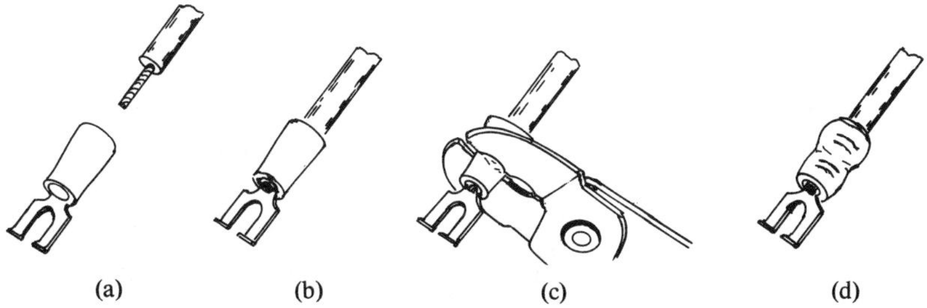

Figure 8–10 Sequence for making a solderless connection.

(cylinder) around the exposed wire. Figure 8–10 shows the proper sequence for making the connection. Add sleeving to the areas of the connection. The cylinder of the connection should be just large enough to accept the bare wire. These connectors are available for almost any wire size and with many types of terminations, such as ring, round, slip on, and flat slip on. The slip-on connectors are available in interlocking styles to make connections that can be easily separated. Solderless connections can also be used instead of wire nuts to secure rattail splices. It is recommended that no more than two wires be secured with this type of connection.

Many connectors come with insulation on the cylinder portion of the connector, eliminating the need to add sleeving. Frequently, shrink tubing is used for added strength at the point of connection. Figure 8–11 shows the proper method for adding shrink tubing.

SPLICING TEST

1. List three reasons for splicing a conductor.
2. List the proper steps for making a tee splice.
3. What is the advantage of using shrink tubing?
4. List the proper sequence for applying shrink tubing to a wire splice.

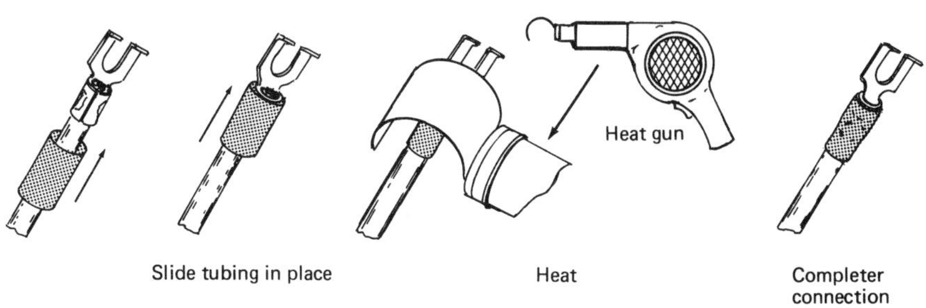

Figure 8–11 Using shrink tubing.

5. Should the three types of splice mentioned in this unit be soldered? Why?
6. What is the proper sequence for crimping a solderless connector?
7. Where are wire nuts used? What type of splice are they used with?
8. Give one major reason for using a solderless connector on a conductor instead of soldering it to a terminal.
9. What is the major difference between a twisted pair splice and a tee splice?
10. Why is shrink tubing added to a connector other than as an insulator?

LAB PRACTICE: SPLICING AND CRIMPING CONDUCTORS

Materials

- 8-inch no. 18 gauge wire
- 8-inch no. 20 gauge wire
- Solderless terminal lug for no. 20 gauge wire
- 20-inch no. 18 gauge wire (cut in half)
- Shrink tubing
- Wire strippers
- Solder
- Soldering iron
- Heat gun
- Crimping tool

Procedure

Form two splices: A tee splice using the 8-inch no. 20 and no. 18 gauge wire, and a rattail using the 20 inches of no. 18 gauge wire. Follow the procedure shown in Figs. 8-6 and 8-4. Connect the solderless connector to one wire of the rattail splice. Follow the procedures in Fig. 8–10. Add shrink tubing to cover the solderless connection. Follow the procedure in Fig. 8–11.

Appendix A

RESISTER COLOR CODES

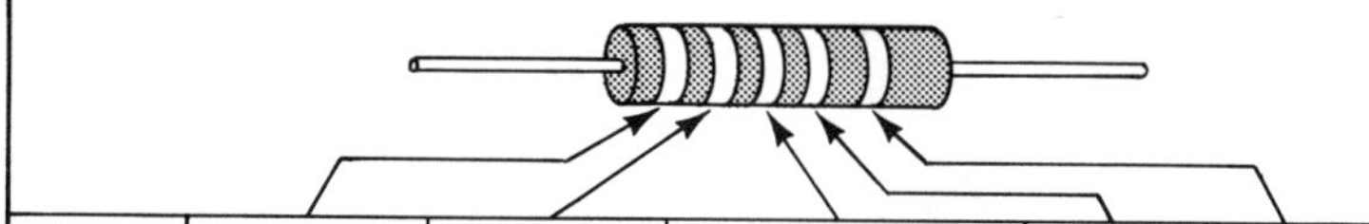

Color	First significant figure	Second significant figure	Multiplier	Tolerance	Relativity level
Black	0	0	1 (10^0)		–
Brown	1	1	10 (10^1)		M 1.0%
Red	2	2	100 (10^2)		P 0.1%
Orange	3	3	1,000 (10^3)		R 0.01%
Yellow	4	4	10,000 (10^4)		S 0.0001%
Green	5	5	100,000 (10^5)		
Blue	6	6	1,000,000 (10^6)		
Violet	7	7	10,000,000 (10^7)		
Gray	8	8	100,000,000 (10^8)		
White	9	9	1,000,000,000 (10^9)		
Gold	–	–	0.1	± 5%	
Silver	–	–	0.01	± 10%	
None	–	–	–	± 20%	

Appendix B

COMMON SOLDER FLUX CONTENT

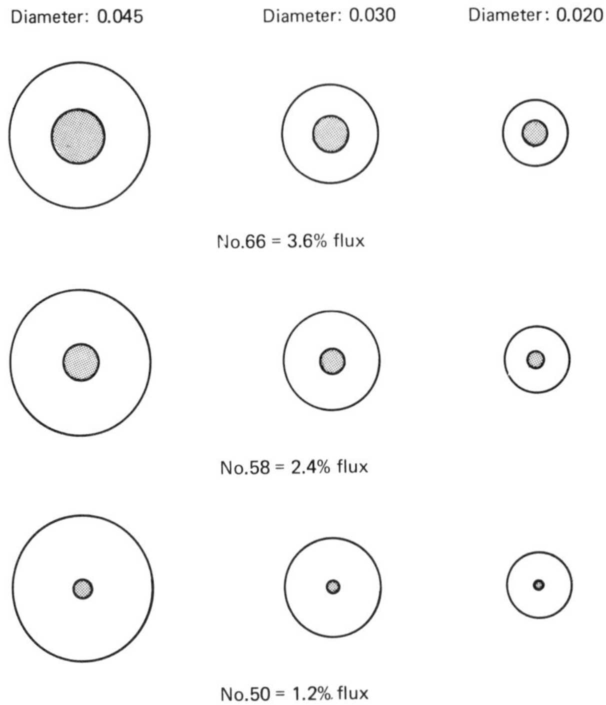

Appendix C

FLUX TYPES

To remove nonmetalic oxide; use:

1. Acid or chloride flux
   Inorganic salts
2. Organic flux
   Fatty acids
3. Rosin flux
   Non corrosive
   Nonconductive
   Resists humidity
   Active hot
   Inactive cold

Appendix D

PROPER NUMBER OF WRAPS FOR MODIFIED AND STANDARD WRAPS

Wire size			Minimum number of turns	
AWG	Diameter in inches	Diameter in mm	Class 2 (modified connection)	Class 1 (standard connection)
16	0.051	1.30	4 stripped and $\frac{1}{2}$ insulated	4 stripped
18	0.0403	1.00	4 stripped and $\frac{1}{2}$ insulated	4 stripped
20	0.032	0.80	4 stripped and $\frac{1}{2}$ insulated	4 stripped
22	0.0253	0.65	5 stripped and $\frac{1}{2}$ insulated	5 stripped
23	0.0226	0.574	5 stripped and $\frac{1}{2}$ insulated	5 stripped
24	0.0201	0.50	5 stripped and $\frac{1}{2}$ insulated	5 stripped
25	0.0179	0.455	$5\frac{1}{2}$ stripped and $\frac{1}{2}$ insulated	$5\frac{1}{2}$ stripped
26	0.0159	0.40	6 stripped and $\frac{1}{2}$ insulated	6 stripped
27	0.0142	0.361	$6\frac{1}{2}$ stripped and $\frac{1}{2}$ insulated	$6\frac{1}{2}$ stripped
28	0.0126	0.32	7 stripped and $\frac{1}{2}$ insulated	7 stripped
29	0.0113	0.287	7 stripped and $\frac{1}{2}$ insulated	7 stripped
30	0.0100	0.25	$7\frac{1}{4}$ stripped and $\frac{1}{2}$ insulated	7 stripped

Appendix E

SOLDERLESS WRAPPING BITS AND SLEEVES

(Courtesy of Cooper Tools, Inc.)

SOLDERLESS WRAPPING BITS AND SLEEVES

AWG = American Wire Gage M = Modified Wrap S = Standard Wrap S.F. = Strip Force B.F. = Back Force A = Alloy Wire
LBS. = Pounds KG = Kilograms G = Grams R.H. = Right Hand L.H. = Left Hand CS&W = Cut, Strip, and Wrap ∇ = Made to Order

(1) Wire Size			(3) Wrap Type	Bit Part No.	N/S No.	Sleeve Part No.	N/S No.
AWG	in.	mm					
18	.040	1.02	S	504222	95210	18285	95153
			M	504908	95212	18640	95159
20	.032	0.81	S	507356	95249	18840	95159
			M	26495	95172	26245	95170
			S	18633	95155	26245	95170
			S	18637	95157	18640	95159
20-22			M	519904	95388	18640	95159
22	.025	.064	S	505413	95223	18840	95159
			M	504939	95214	507939	95258
			M	26498	95173	18640	95159
			M	26699	95175	18640	95159
			S	37014	95180	18840	95160
			S	18632	95154	18840	95160
			S	18635	95156	18640	95159
22-24			S	500131	95183	18840	95160
24	.020	0.51	S	502134	95203	502129	95202
			M	505415	95225	502129	95202
			M	506991	95244	506999	95245
			M	504155	95209	18840	95159
			M	26263	95171	18840	95159
			S	512058	95333	512056	95331
			S	37013	95179	512056	95331
			S	26589	95174	512056	95331
			S	509301	95282	502129	95202
			S	20420	95162	512056	95331
24-26			M	519070	95387	18840	95159
26	.016	0.41	S	505279	95220	507100*	95248
			M	506445	96092	507100*	95248
			M	506445-V	95233	507100*	95248
			M	506420	95232	502129	95202
			M	509405	95286	502129	95202
			M	511250	95306	501098	95188
			M	502118	95201	512056	95331
			S	504910	95213	512056	95331
			M	517104	95361	512056	95331
			M	506781	95243	512056	95331
28	.013	0.33	M	509278	96091	507100*	95248
			M	509278-V	95270	507100*	95248
			M	501389	95193	502129	95202
			S	508748	95261	507100*	95248
			S	506633	95242	507100*	95248
			M	505412	95222	512056	95331
			S	505373	95221	512056	95331
			M	511304	95311	512056	95331
30	.010	0.25	M	501097	95187	507100*	95248
			S	500352	95185	507100*	95248
			M	511208	95303	507100*	95248
			M	507063	96090	507100*	95248
			M	507063-V	95247	507100*	95248
			M	511440	95318	507100*	95248
			M	519936	95396	507100*	95248
			S	500353	95186	507100*	95248
			M	507573	95254	507100*	95248
			M	507502	95254	507100*	95248
			M	501381	95192	512056	95331
			M	501194	95190	502129	95202

A = Alloy wire O = OFHC wire A° = Alloy if terminal diagonal at low end
* = Cat. No. 507100-V N/S No. 95248 for Carded Version. O° = OFHC if terminal diagonal at low end.

SOLDERLESS WRAPPING BITS AND SLEEVES (cont.)

O = OFHC (Oxygen Free Copper Wire) R = Radius IN = Inches MM = Milimeters O.D. = Outside Diameter I.D. = Inside Diameter DIA. = Diameter Min. = Minimum Max. = Maximum Sq. = Square WT = Wall Thickness T = Thickness L = Length W = Width

(2) Wire Type	(6) Terminal Diagonal Minimum	Maximum	(7) Terminal Hole Depth	(4) Maximum Insulation Diameter	Minimum No. of Turns of Insulation	Effective Radius
	"/mm	"/mm	"/mm	"/mm		"/mm
A, 0	.061/1.56	.074/1.88	1.00/25.4			.150/3.81
A, O	.061/1.56	.074/1.88	1.00/25.4	0.70/1.78	1	.151/3.84
A, O	.042/1.07	.074/1.88	1.00/25.4			.121/3.07
A, O	.042/1.07	.074/1.88	1.00/25.4	.059/1.50	9/10	.150/3.81
A, O	.059/1.50	.091/2.31	1.00/25.4			.147/3.73
A, O	.091/2.31	.123/3.12	1.00/25.4			.150/3.81
A, O	.054/1.37	.069/1.75	1.00/25.4	.070/1.78	3/4	.142/3.61
A, O	.049/1.24	.074/1.88	1.00/25.4			.119/3.02
A, O	.049/1.24	.074/1.88	1.00/25.4	.052/1.32	1	.131/3.33
A, O	.049/1.24	.074/1.88	1.00/25.4	.059/1.45	1	.135/3.43
A, O	.049/1.24	.074/1.88	1.00/25.4	.059/1.45	1-1/4	.135/3.43
A, O	.061/1.55	.076/1.93	1.00/25.4			.125/3.17
A, O	.061/1.55	.066/2.18	1.00/25.4			.125/3.17
A, O	.098/2.49	.123/3.12	1.00/25.4			.142/3.61
A, O	.054/1.37	.074/1.88	1.00/25.4			.112/2.84
A, O	.024/0.61	.044/1.12	1.00/25.4			.084/2.13
O	.024/0.61	.044/1.12	1.00/25.4	.044/1.12	1	.098/2.49
A	.047/1.19	.067/1.70	1.00/25.4	.041/1.04	9/10	.100/2.54
O	.049/1.24	.069/1.75	1.13/28.5	.050/1.27	9/10	.119/3.02
A, O	.054/1.33	.074/1.88	1.13/28.8	.046/1.17	9/10	.117/2.97
A, O	.055/1.40	.074/1.88	1.00/25.4			.105/2.67
A, O	.055/1.40	.074/1.88	1.00/25.4			.105/2.67
A, O	.055/1.40	.074/1.88	1.50/38.1			.114/2.90
A, O	.059/1.50	.067/1.70	1.00/25.4			.087/2.21
A, O	.073/1.85	.092/2.34	1.00/25.4			.105/2.67
A, O	.054/1.37	.074/1.88	1.13/28.8	.046/1.17	9/10	.117/2.97
A, O	.023/0.58	.039/0.99	0.75/19.1			.068/1.73
A, O	.023/0.58	.039/0.99	0.75/19.1	.031/.079	1-1/4	.074/1.88
A, O	.023/0.58	.039/0.99	0.75/19.1	.031/.079	1-1/4	.074/1.88
A, O	.028/0.71	.044/1.12	0.75/19.1	.044/1.12	1-1/8	.098/2.49
A, O	.034/0.86	.051/1.29	1.00/25.4	.034/0.86	1	.086/2.18
O	.031/0.79	.036/0.91	0.75/19.1	.043/1.09	1	.097/2.46
A, O	.053/1.35	.069/1.75	1.00/25.4	.041/1.04	9/10	.111/2.82
A, O	.058/1.47	.074/1.88	1.00/25.4			.105/2.67
O	.054/1.33	.058/1.47	1.00/25.4	.041/1.05	1	.100/2.54
A, O	.059/1.50	.069/1.75	1.00/25.4	.041/1.04	9/10	.111/2.82
O	.031/0.79	.034/0.86	0.75/19.1	.030/0.76	9/10	.067/1.70
O	.031/0.79	.034/0.86	0.75/19.1	.030/0.76	9/10	.067/1.70
A°, O°	.033/0.84	.036/0.91	0.75/19.1	.036/0.91	9/10	.087/2.21
A, O	.034/0.86	.037/0.94	0.75/19.1			.063/1.60
O	.041/1.04	.044/1.12	0.75/19.1			.066/1.68
A, O	.056/1.42	.063/1.60	1.00/25.4	.035/0.89	9/10	.104/2.64
A, O°	.066/1.68	.069/1.75	1.00/25.4			.105/2.67
O	.066/1.68	.069/1.75	1.00/25.4	.038/0.97	9/10	.100/2.54
A, O	.026/0.66	.031/0.79	0.75/19.1	.027/0.69	9/10	.068/1.73
O	.029/0.74	.031/0.79	0.75/19.1			.067/1.70
O	.031/0.79	.034/0.86	0.75/19.1	.023/0.58	1-1/2	.061/1.55
O	.031/0.79	.034/0.86	0.75/19.1	.023/0.58	9/10	.061/1.55
O	.031/0.79	.034/0.86	0.75/19.1	.023/0.58	9/10	.061/1.55
O	.031/0.79	.034/0.86	0.75/19.1	.027/0.69	1-1/8	.064/1.63
A	.031/0.79	.038/.097	0.75/19.1	.028/0.71	1	.068/1.73
A°, O	.032/0.81	.035/0.89	0.75/19.1			.061/1.55
A	.033/0.84	.036/0.91	0.75/19.1	.023/0.58	1-1/4	.061/1.55
A°, O°	.041/1.04	.044/1.12	0.75/19.1	.027/0.69	1-1/4	.071/1.80
A	.060/1.52	.063/1.60	1.00/25.4	.027/0.69	1	.107/2.72
A	.066/1.68	.069/1.75	1.00/25.4	.027/0.69	9/10	.089/2.26

NOTE: Tin-plated wire should be considered as Alloy in bit selection.

Appendix F

COPPER WIRE DATA

AWG B & S gauge	Turns per linear inch: Enamel	DCC	SCC	Nylon	Diameter (inches)	Circular mils	Current-carrying capacity at 700 CM per amp	Feet per lb (Bare)	Ohms per 1,000 ft at 20°C
0000	—	—	—	—	0.4600	211,600	302.3	1.561	0.04901
000	—	—	—	—	0.4096	167,800	239.7	1.968	0.06180
00	—	—	—	—	0.3648	133,100	190.1	2.482	0.07793
0	—	—	—	—	0.3249	105,500	150.7	3.130	0.09827
1	—	3.3	3.3	—	0.2893	83,690	119.6	3.947	0.1239
2	—	3.6	3.8	—	0.2576	66,370	94.8	4.977	0.1563
3	—	4.0	4.2	—	0.2294	52,640	75.2	6.276	0.1970
4	—	4.5	4.7	—	0.2043	41,740	59.6	7.914	0.2485
5	—	5.0	5.2	—	0.1819	33,100	47.3	9.980	0.3133
6	—	5.6	5.9	—	0.1620	26,250	37.5	12.58	0.3951
7	—	6.2	6.5	—	0.1443	20,820	29.7	15.87	0.4982
8	7.6	7.1	7.4	—	0.1285	16,510	23.6	20.01	0.6282
9	8.6	7.8	8.2	—	0.1144	13,090	18.7	25.23	0.7921
10	9.6	8.9	9.3	—	0.1019	10,380	14.8	31.82	0.9989
11	10.7	9.8	10.3	—	0.09074	8,234	11.8	40.12	1.260
12	12.0	10.9	11.5	—	0.08081	6,530	9.33	50.59	1.588
13	13.5	12.0	12.8	—	0.07196	5,178	7.40	63.80	2.003
14	15.0	13.8	14.2	14.9	0.06408	4,107	5.87	80.44	2.525
15	16.8	14.7	15.8	—	0.05707	3,257	4.65	101.4	3.184
16	18.9	16.4	17.9	18.6	0.05082	2,583	3.69	127.9	4.016
17	21.2	18.1	19.9	—	0.04526	2,048	2.93	161.3	5.064
18	23.6	19.8	22.0	23.2	0.04030	1,624	2.32	203.4	6.385
19	26.4	21.8	24.4	—	0.03589	1,288	1.84	256.5	8.051
20	29.4	23.8	27.0	28.9	0.03196	1,022	1.46	323.4	10.15
21	33.1	26.0	29.8	—	0.02846	810.1	1.16	407.8	12.80
22	37.0	30.0	34.1	36.0	0.02535	642.4	0.918	514.2	16.14
23	41.3	31.6	37.6	—	0.02257	509.5	0.728	648.4	20.36
24	46.3	35.6	41.5	44.7	0.02010	404.0	0.577	817.7	25.67
25	51.7	38.6	45.6	—	0.01790	320.4	0.458	1,031	32.37
26	58.0	41.8	50.2	55.7	0.01594	254.1	0.363	1,300	40.81
27	64.9	45.0	55.0	—	0.01420	201.5	0.288	1,639	51.47
28	72.7	48.5	60.2	69.4	0.01264	159.8	0.228	2,067	64.90
29	81.6	51.8	65.4	—	0.01126	126.7	0.181	2,607	81.83
30	90.5	55.5	71.5	86.2	0.01003	100.5	0.144	3,287	103.2
31	101.0	59.2	77.5	—	0.008928	79.70	0.114	4,145	130.1
32	113.0	62.6	83.6	106.0	0.007950	63.21	0.090	5,227	164.1
33	127.0	66.3	90.3	—	0.007080	50.13	0.072	6,591	206.9

(continued)

COPPER WIRE DATA *(continued)*

AWG B & S gauge	Turns per linear inch				Diameter (inches)	Circular mils	Current-carrying capacity at 700 CM per amp	Feet per lb (Bare)	Ohms per 1,000 ft at 20°C
	Enamel	DCC	SCC	Nylon					
34	143.0	70.0	97.0	133.0	0.006305	39.75	0.057	8,310	260.9
35	158.0	73.5	104.0	—	0.005615	31.52	0.045	10,480	329.0
36	175.0	77.0	111.0	167.0	0.005000	25.00	0.036	13,210	414.8
37	198.0	80.3	118.0	—	0.004453	19.83	0.028	16,660	523.1
38	224.0	83.6	126.0	206.0	0.003965	15.72	0.022	21,010	659.6
39	248.0	86.6	133.0	—	0.003531	12.47	0.018	26,500	831.8
40	282.0	89.7	140.0	263.0	0.003145	9.89	0.014	33,410	1,049.0

Appendix G

POOR SOLDER JOINTS AND CONNECTIONS

Defect	Description	Cause	Solution
		Joints	
Cold joint	Dull, grayish appearance	Not enough heat Poor contact between iron and joint	Remelt the joint
Hot joint	Grayish granular appearance, sometimes with pinholes	Too much heat	Add flux and remelt the joint
Excessive solder	Buildup of solder; cannot see contour of wire	Too much solder	Remove excessive solder
Insufficient solder	Amount of solder is not sufficient to provide a good solder fillet	Not enough solder	Reheat and add more solder
Disturbed or fractured joint	Irregular crystallized appearance, cracks or cleavagelike lines	Wire or component was moved while the solder was cooling	Remelt the solder and hold the part still until it solidifies
Poor mechanical connection	Weak connection sometimes causes fractured joints or excessive solder	Wire or lead not tightly wrapped around the terminal	Redo it
Exposed copper	Copper exposed	Wire or lead is scratched or cut	Tin the copper with solder
Rosin joint	Flux trapped between parts being soldered	Insufficient heat or improper cleaning of the parts	Take the joint apart, clean the surface, and resolder it
Solder projection	Solder sticks out of connection	Too much heat; icicles are caused by removing the iron from the connection too slowly	Remove all the excessive solder and add a drop of flux; then reheat the joint

(continued)

POOR SOLDER JOINTS AND CONNECTIONS *(continued)*

Defect	Description	Cause	Solution
	Stranded Wire Connections		
Wicking	Insulation appears melted from the inside	Too much heat	Redo it
Burned insulation	Insulation exposes the conductor	Insulation susceptible to heat	Replace the wire
Birdcaging	Separated wires	Poor tinning or flexing of wire	Replace the wire
Nicked or cut strands	Conductors are damaged	Wrong gauge used to strip the wire	Replace the wire
Improper strip gap	Insulation gap too long or too short	Poor measurement of the gap	Replace the wire

Appendix H

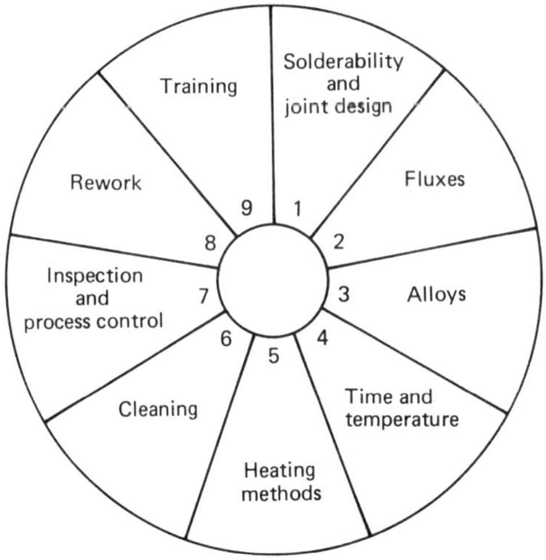

ELEMENTS OF SOLDERING

To obtain good soldered joints, nine key elements must be considered in the soldering process.

1. *Solderability and joint design:* Surfaces of the components and printed wire boards must be clean. The joint must be mechanically stable and dimensioned to promote solder flow across the surfaces being joined.
2. *Fluxes:* Surfaces are coated with flux to act as a cleaning agent and to prevent surface oxidation during the high-temperature phase when the solder is applied.
3. *Alloys:* Solders are metal alloys, principally alloys of tin and lead. The relative proportion of tin to lead gives each solder type specific characteristics.
4. *Time and temperature:* For the solder to flow correctly over the surfaces, needed to produce a high solder joint, the solder and surfaces being joined must be raised to the correct temperature.
5. *Heating methods:* The application of heat (and solder) takes many forms. Hand soldering and machine (wave) soldering are the two most common methods.
6. *Cleaning:* Residue flux can be removed after the soldering operation. The degree of cleaning required depends on the type of flux employed and the end use of the product.

7. *Inspection and process control:* There are many points within the soldering process at which inspection is applied. The degree of inspection will vary and will reduce as confidence in the process increases. Both preventive and also corrective actions must be taken.
8. *Rework:* A printed wire board may require additional work in the form of correcting minor solder faults, adding sensitive components, or replacing faulty components isolated at board test.
9. *Training:* Operators for all stages of the soldering process must receive suitable training if they are to perform their tasks correctly.

Appendix I

STANDARD REFERENCES AND UNITS OF MEASURE

Components	Unit of measure	Designator Number
Resistors	Ohms (Ω)	R
Capacitor	Farad (F)	C
Inductors	Henry (H)	L
Diode		CR
Zener diode		VR
Transistor		Q
Power transformer		TR
FETs		Q
IC		U
Transformer	Henry (H)	T
Op amp		AR
Switch		SW
Plug-removable connector		P
Jack-fixed connector or receptacle		J
Blower fan		BL
Terminal strip		E
Thermocouple (measures heat)		TC
Fuse number		F
Edge connector		X, XL
Solid-state devices:		
Filters		FL
Isolators/pad/antenuator		AT
Mixer		A
Oscillator		OS
Power supply		PS

Appendix J

GENERAL SHOP RULES: SAFETY AND PRACTICE

1. Always obey all posted warning signs. They are posted for your safety.

2. Caution anyone you see violating a safety rule.

3. Work at a speed consistent with safety. Don't take risks. Foolishly hurrying, to finish a project or task, is dangerous.

4. Never work on electrical equipment alone.

5. Never work on live equipment when fatigued and always keep one hand in a pocket when working on live equipment (a current through one side of the body can cause severe burns, but seldom death).

6. Do not touch electrical equipment while standing on either a damp or metal floor or while either you or the equipment is wet or damp.

7. Do not handle electrical equipment that is not grounded.

8. Do not use water on an electrical fire.

9. Remove a victim from an electrical circuit as quickly as possible without endangering your safety (use a dry piece of wood, cloth, or other nonconducting material). Any delay may prove fatal.

10. Start artificial resuscitation on an electrical shock victim at once and continue until a medical expert arrives.

11. Never carry bulky objects that obstruct vision.

12. Grasp the plug, not the cord, to remove the line cord from an electrical outlet.

13. Wear protective clothing while working in the lab.

14. Wear short-sleeved clothing around drill presses and other revolving machines. Do not wear neckties, rings, bracelets, brooches, or anything that can become entangled in mechanical equipment.

15. Keep long hair under a hat or tied back while you work. Long hair can easily be caught in revolving machinery, causing severe scalp laceration.

16. Never allow anyone to turn power on or off for you while working on electrical equipment or machinery.

17. Know the locations of emergency exits, electrical switches, and fire extinguishers.

18. When lifting heavy objects, keep your back and arms as straight as possible, bend your knees, and lift with your leg muscles.

19. Never use flammable liquids, such as paint thinners, naphtha, and gasoline, in enclosed areas.

20. Good housekeeping is an important factor in good safety at all times:

a. Clean up immediately any liquids spilled on workbench or floor.

b. Keep your work area neat and clean. Put waste material into the proper receptacles.

c. Keep floors and passageways clear of equipment, electrical cords, and all materials.

d. Keep tools and materials from projecting over the edges of benches and tables whenever possible.

21. Keep all assembly areas well ventilated; fumes from lead in the solder are a hazard to your health.

22. Know the company evacuation plan for fires, earthquakes, or other catastrophes.

23. *Observe safety practices* for yourself, for others, for equipment.

CLASS A FIRES

Fires involving combustible material, such as wood, cloth, paper, rubber, and many plastics. To extinguish, use water to cool and quench below the ignition temperature. Soda-acid (CO_2, carbon dioxide) extinguishers may also be used.

CLASS B FIRES

Fires involving flammable liquids, such as gasoline, kerosene, greases, and gases. The smothering or blanketing effect of oxygen exclusion is most effective. Foam and CO_2 types of extinguishers may be used.

CLASS C FIRES

Fires involving energized electrical equipment. Use a nonconducting type of extinguishing agent, such as CO_2 or dry powder. It is best to attempt to turn off the power (switch) and then treat as a class A or B fire.

CLASS D FIRES

Fires involving combustible metals, such as magnesium, titanium, and potassium. The extremely high temperature of these types of burning metals makes water and other common extinguishing agents ineffective. Special extinguishing agents are used specifically for each type of combustible metal and are marked specifically for that metal.

No matter what type of fire you encounter, in all cases, point the fire extinguisher nozzle at the base of the fire and not at the top of the flames.

SAFETY TEST

This test should be given to every employee or student who works in an electronics lab. Correct answers should be required before being allowed to work in the laboratory.

Write the correct response in the space provided before each question.

_____ **1.** The floor, aisles, and passageways should be kept clear of stock, tool, and materials. Objects on the floor:

(a) may be left there if the operator of the machine is in a hurry.
(b) may cause someone to slip or trip into a moving machine.
(c) may be ignored at times.
(d) are unsightly.

_____ **2.** Students must not talk to or distract a person operating a machine because:

(a) the operator is likely to be injured.
(b) conversation slows down the flow of work.
(c) the operator is likely to make a mistake.
(d) conversation is annoying to the operator.

_____ **3.** Long hair is dangerous around shop equipment. If it is long enough to get caught in the machine, it must be:

(a) tied up and back.
(b) burned off.
(c) pulled out.
(d) none of the above.

_____ **4.** Loose clothing must be securely fastened or removed and long, loose sleeves rolled up above the elbows:

(a) before operating any machinery.
(b) after operating.
(c) during the operation of a machine.

(d) only when you are assisting the teacher.

_____ 5. When touching electrical switches, plugs, or receptacles, be sure your hands are dry because:

(a) a switch will not operate properly if your hands are wet.

(b) a plug will easily slip from your fingers if your hands are wet.

(c) if your hands are wet, you may receive a severe shock and serious burns.

(d) none of the above.

_____ 6. Water should never be used to put out what kind of fires?

(a) Wood fires

(b) Electrical and oil fires

(c) Paper fires

(d) None of the above

_____ 7. The proper procedure to fight a fire with a fire extinguisher is to:

(a) point the nozzle at the top of the fire.

(b) point the nozzle at the middle of the flame.

(c) cover the area around the fire and keep it from spreading.

(d) point the nozzle at the source of the fire because that is where the fire is located.

The following are true–false questions. Circle T for true and F for False.

T F 1. You should never work on energized equipment alone.

T F 2. You may be severely shocked (complete an electrical circuit through your body) if you stand on a damp or metal floor while working on an electrical device.

T F 3. For safety, all electrical equipment should be grounded.

T F 4. Water should be used on electrical fires.

T F 5. An electrical shock victim should be pulled away from the circuit without touching his or her body.

T F 6. Artificial resuscitation should be started at once for an electrical shock victim.

T F 7. Eye protection is optional when soldering above eye level.

T F 8. Good housekeeping is part of good safety practices.

T F 9. Soldering irons should be picked up by their handles even if you think they are cold.

T F 10. Wires being unsoldered can cause solder to spray and cause burns.

T F 11. You should make high-voltage adjustments by placing both hands on the equipment.

T F 12. Burning insulation such as Teflon produces poisonous gases.

T F 13. Static-sensitive components can be handled away from the static-free work station if care is used.

T F 14. There must be a common ground at each work station.

Glossary

AC: Alternating current. Electrons flow in one direction, stop, and then flow in the opposite direction. AC can be thought of as electrons moving back and forth, varying in amplitude from a maximum to a minimum value.

Acid brush: Hair or synthetic brush used to clean rosin from a completed solder joint or tinned wire.

Air movement: Force (energy) caused by the movement of air under pressure.

Alloy: Composition of two or more elements, such as solder, a tin–lead alloy.

Analog meter: Meters that use meter movements (needle and dial) for reading a measured value.

Anchor: To secure firmly.

Artificial resuscitation: Method of reviving a person who has stopped breathing, due to electrical shock, drowning, or strangling. Breathing is started by manipulating the diaphragm so as to force air into the lungs or by breathing air through the mouth.

AWG: American Wire Gage, a standard numerical coding system used for measuring conductor width (without insulation).

Bifurcated (split) terminal: Terminal containing a slot or split in which wires or leads are placed before soldering. It is used to provide more than one point of entry.

Birdcaging: Separation of the strands of wire on a conductor.

Braid: Woven, stranded copper wire, used to remove solder from a component or printed circuit pad and trace.

Breakout: Exit point of a number of conductors along various points of a main cable (bundle of wires). This point is usually harnessed (tied) with nylon straps or twine saturated with wax to prevent slippage.

Bridging: Buildup of a solder coating between parts, parts leads, and/or conductor paths, forming an elevated path (*see* fillet).

Capacitance: Property of a capacitor that gives it the ability to store energy.

Capillary action: Force that raises part of the surface of a liquid that is in contact with a solid. The ability of a porous surface to soak up a liquid.

Cavitation: Formation of partial vacuums in a liquid by a swiftly moving solid body or by high-frequency sound waves.

Cell: Single unit of a battery that produces a direct voltage by converting chemical energy into electrical energy.

Chassis: Housing for electrical devices and equipments.

Cold solder joint: Unsatisfactory solder connection resulting from dewetting or movement of the conductor during cooling and frequently exhibiting an abrupt rise of the solder from the surface being soldered. These usually appear frosty and granular.

Conductor: Lead or wire, solid or stranded, or printed wiring path serving as an electrical interconnection between terminations.

Connection: Electrical termination. The joining point of two or more wires or components.

Cored solder: Solder formed into a wire manufactured with rosin flux in its center. Flux is automatically applied as the solder melts.

Cracked solder joint: Soldered connection that has fractured or broken within the solder.

Crimping: Process of securing or clamping a connector to a conductor by mechanical means.

Current: Movement of electrons through a wire or conductor, measured in amps.

Dewetting: Condition in a soldered area in which the liquid solder has not bonded chemically; characterized by an abrupt boundary between solder and conductor or solder and terminal or termination area.

Diameter: Thickness of a material such as wire, given in inches, centimeters, or mils (0.001 inch).

Dielectric: Insulating material between two plates of a capacitor: air, paper, oil, mica, or ceramic.

Digital meter: Meter that uses LED numerical displays for indicating a measured value.

Disturbed solder joint: Unsatisfactory connection resulting from relative motion between the conductor and termination during solidification of the solder.

Dress: Refers to the neatness of the insulation, the length of the wire exposed, and the form of the wire, that is, the overall exactness of the wiring.

Electrical circuit: Number of conductors connected together in order to carry an electrical current, or the interconnection of devices in one or more closed paths to perform an electronic function.

Electrical component: Assembly of one or more electronic or electrical parts that may be disassembled or separated without destruction of designed use (printed wiring assembly).

Electrical impulse: Sudden change in the level of current or voltage.

Electrical shock: Flow of an electric current through the body. Electrical shock can cause muscle twitching or paralysis, burns, interruption of breathing, unconsciousness, ventricular fibrillation, cardiac arrest, or death.

Electrical wave: Wave that comprises electric and magnetic field energy traveling through space with the speed of light.

Electrodes: Conducting elements that serve to heat objects such as thermal strippers.

Electromagnetic induction: Voltage produced in a coil as the number of magnetic lines of force passing through a coil changes.

Ethyl alcohol: Liquid used for cleaning components as preparation for soldering them.

Excessive solder joints: Unsatisfactory connection wherein the solder obscures the configuration of the connection.

Eyelet: Tubular metal part having both ends headed or rolled over.

Fillet: Smoothed, concave buildup of material between two surfaces; solder between a part lead and a solder pad or terminal, or a coating of solder between a part and a PC board.

Flush: Even, level.

Flux: Material made of rosin; used to remove nonmetallic oxides to help improve the soldering process.

Flux residue: Result of heat applied to the solder and flux, causing the flux to bubble, spread, and solidify. Flux residue should be completely removed after the soldering process is complete.

Fractured solder joint: Joint that shows evidence of cracking. Essentially the same as a disturbed joint.

Frequency: Number of ac cycles in a second, measured in hertz. The ac voltage coming into our homes is 60 hertz or 60 cycles per second.

Grounded: Metallic connection that has the same voltage level as earth and can have an actual connection to it. The metal case of a piece of equipment is usually grounded by this method.

Harness board: Template used for production harnessing.

Heat gun: Hair blower-like tool used to heat shrink tubing.

Heat sink: Device used to absorb or dissipate excess heat away from heat-sensitive devices and wires.

Hook terminal: Terminal formed in a hook shape.

Idling temperature: Temperature of the soldering iron while not in use.

Impregnate: To fill or saturate.

Inductance: Property of an inductor that allows the coil to store energy and causes it to oppose any change of current.

Induction (static): Charge developed on a conductor as a result of its proximity to a charged body.

Inert: Exhibiting no chemical activity.

Insulation: Protective covering on a conductor or wire.

Joint: Solder joint or termination (connection point).

Jumper lead: Short wire used to temporarily complete a circuit or to bypass a circuit.

Junction box: In electrical wiring, a box for joining different runs of wiring or cabling.

Lacquer thinner: Solvent to thin or remove lacquer.

Lacing: Drawing together by means of ties or laces; tying together.

Land area: Pad or area to which a surface-mounted component is adhered.

LED: Light-emitting diode.

Leverage: Increase of mechanical force; a wrench with a long handle (increased leverage) provides added torque at the nut.

Liquefy: To melt; to make solder flow.

Magnetic field: Invisible field around a magnet or wire through which current is passing.

Metal solvent action: Forming of a new alloy by two or more different metals when heated to their melting point.

Metallurgical: Science and technology of metals.

Micro: Prefix meaning 1×10^{-6}, or 0.000001.

Mockup: Full-size model built to scale. In electronics it is used to build a template for a harness board.

Molecule: Smallest particle of a substance that retains the physical properties of the substance and is composed of two or more different atoms.

Molten: Fused or liquefied by heat.

Oscillator: Electronic device that converts dc energy into ac energy. It generates ac power at a single predetermined frequency.

Oscilloscope: Electronic instrument that displays ac wave forms on the screen of a cathode-ray-tube.

Oxide: Compound that forms on the surface of an element when it has been exposed to air.

Pad: Termination point on a printed circuit board.

Part lead: Conductor, solid or stranded, attached to a part.

Pico: Prefix meaning 1×10^{-12} or 0.000000000001.

Pigtail: Small wire extending from either a component or a connection.

Pit: Small holes or sharp depressions in the surface of the solder.

Plated-through hole (PTH): Hole formed by a layer of copper material on the inside surface of the hole. It is also known as a support hole. The PTH is used to provide additional mechanical strength to the soldered termination or to provide an electrical interconnection on a multilayer printed circuit board.

Post: Stud or terminal. An upright post used as an anchor for wires.

Primary winding: In a transformer, it is the coil winding that sets up (passes) a current to one or more secondary windings.

Printed circuit board: Insulating board onto which a circuit has been printed.

RAM: Random access memory. Also called read-write memory (your work area). Lets the computer store and retrieve information. You can erase the contents of read-write memory to make room for new information.

Reference letter: Letter on a schematic drawing that represents a component. For instance the letter R is used for resistors.

Reflow soldering: Components are precoated with solder and are joined by the application of heat to reflow the solder.

Resistance: Property of any substance or material (including the human body) that limits or controls the passage of electrical current.

Rework: Repairing or correcting soldered material to make it conform to specifications.

Run: Continuous connection of electrical points in a circuit that are common to each other.

Secondary winding: Output winding of a transformer. The output current is due to inductive coupling with another coil called the primary.

Service loop: Length of extra wire that is left on each conductor so that the conductor (wire) may be removed for service and resoldered to the terminal.

Shrink tubing: Strong plastic tubing that shrinks when heat is applied.

Solder cup terminal: Hollow, cylindrical terminal that can hold one or more conductors.

Solder icicle: Cone-shaped peak or sharp point of solder usually formed by removing the soldering iron from the connection too slowly.

Solder pool: Small drop of solder used to help transfer heat to items being soldered.

Solder pot: Container that holds molten solder.

Solidifies: To make or become solid; when solder cools, it solidifies.

Solvent: Chemical cleaner used to remove rosin, oil, and dirt from the surface of items to be soldered.

Sound waves: Alteration in pressure, stress, particle displacement or velocity, and so on, which is reproduced in an elastic material. Progressive vibration by which sounds are transmitted.

Specification: Accompanying paper or document that details information about equipment, materials, or processes to be followed.

Static electricity: Electricity at rest.

Stress relief: Method or means to minimize stresses to the soldered termination or part.

Stressed: Pressure or strain that tends to distort an object.

Strip gap: Distance of the exposed wire (area where insulation has been removed) between the terminal and the beginning of the exposed wire.

Surface mount: Components mounted on the surface of a PC board rather than using plated-through holes in the board.

Tapped: Tapered and threaded.

Template: Gauge or pattern used as a guide for forming similar items.

Terminal: Tie point at which electrical conductors are joined.

Termination: Point at which electrical conductors or connectors are joined.

Tinning: Coating of a surface with a uniform layer of solder.

Trace: Ribbon of copper conductive material (such as copper) or plating that is on a printed circuit board.

Turret terminal: Round, post-type, grooved stud around which conductors are fastened before soldering.

Watt: Standard unit of electrical power. Most electric companies express watts in terms of power used in one hour. A 60-watt light bulb uses 60 watts of energy in one hour.

Wetting: Flow and adhesion of a liquid to a base metal characterized by smooth, unbroken, even edges.

Wicking: Flow of molten solder or cleaning solution up under the insulation by capillary action.

Wire pinch: Wire being squeezed between the hard surfaces of the cutters, causing the wire to become thin and compressed at the end instead of flat.

Bibliography

ALLEN, B. M., *Soldering Handbook: A Practical Manual for Industry and the Laboratory.* New York: Drake Publications Ltd., 1970.

BILOTTA, ANTHONY, *Connections in Electronic Assemblies.* New York: Marcel Dekker, Inc., 1985.

DUARTE, S. R., and R. L. DUARTE, *Electronics Assembly and Fabrication Methods.* New York: McGraw-Hill Book Co., 1973.

HINCH, STEPHEN W., *Handbook of Surface Mount Technology.* New York: John Wiley & Sons, Inc., 1988.

HOLLOMAN, JAMES K., JR., *Surface-mount Technology for Printed Circuit Board Design.* Indianapolis, Ind.: Howard W. Sams, 1989.

LAMBERT, LEO P., *Soldering for Electronic Assemblers.* New York: Marcel Dekker, Inc., 1988.

MARKO, HOWARD, *Soldering Handbook for Printed Circuit and Surface Mounting.* New York: Van Nostrand Reinhold, 1986.

———, *Solders and Soldering.* New York: McGraw-Hill Book Co., 1986.

SHIMIZU, GORDON, *Electronic Fabrication.* Albany, N.Y.: Delmar Publishers, 1980.

WOODGATE, RALPH W., *The Handbook of Machine Soldering: A Guide for the Soldering of Electronic Printed Wiring Assemblies.* New York: John Wiley & Sons, 1988.

Technical Magazines

JEE Journal of Electronic Engineering, DEMPA Publications, Tokyo.

Surface Mount Technology, Lake Publications Corp., Libertyville, Ill.

Index

C

D

E

F

N

O

P

Q

R

S